PRACTICAL ELECTRICITY

by Robert G. Middleton

THEODORE AUDEL & CO.
a division of

HOWARD W. SAMS & CO., INC.
4300 West 62nd Street
Indianapolis, Indiana 46268

FOURTH EDITION

1978 PRINTING

International Standard Book Number: 0-672-23218-9
Library of Congress Catalog Card Number: 73-94187

Foreword

Although the simplest possible language has been used throughout this book, various technical words and phrases have necessarily been included; these are very carefully explained both in the text and in the glossary. Comprehensive coverage of topics has been accomplished within a book of reasonable size. Considerable emphasis has been placed on functional illustrations, to provide the maximum amount of information.

Great care has been taken to ensure that all material is accurate and that procedures are in accordance with good practice. This new edition places greater emphasis on circuiting and the underlying laws of circuit action. Magnetic circuits are developed to the extent required by the practical electrician in his work. Somewhat less background knowledge is required of the reader than in previous editions, and the text progresses from the most elementary considerations to the comparatively complex situations that must be contended with by the journeyman electrician.

Practical Electricity will find valuable use as a text book in technical institutes, as well as a self-instruction guide for the general reader. Mathematics is avoided in the text wherever possible, although simple arithmetic and geometry is essential in the treatment of DC and AC circuits. When trigonometry is occasionally required in the course of discussion, its basic meaning is fully explained.

Various topics, such as electrical instruments, electrical test procedures, and electrical symbols are fully discussed. Electric circuits are also fully covered, which include series-parallel circuits, two-, three-, and four-wire distribution circuits, and inductive-capacitive AC circuits. Wiring requirements for the home, lighting calculations, electric heating, intercommunication, and alarm installation are also covered.

ROBERT G. MIDDLETON

Contents

CHAPTER 1

Magnetism
and Electricity

Early experimenters dating back to the "dawn of history" discovered that certain hard black stones attracted small pieces of iron. Later, it was also discovered that a *lodestone* or "leading stone" pointed North-and-South when freely suspended on a string, as shown in Fig. 1. Lodestone is a magnetic ore that becomes magnetized if lightning happens to strike nearby. Today, we use magnetized steel needles instead of lodestones in magnetic compasses. Fig. 2 illustrates the appearance of a modern pocket compass.

MAGNETIC POLES

Any magnet has a North and a South pole. We know that the earth is a huge, although weak, magnet. In Fig. 1, the end of the lodestone that points toward the North Star is called its North-seeking pole; the opposite end of the lodestone is called its South-seeking pole. It is a basic law of magnetism that *like poles repel each other* and *unlike poles attract each other*. For example, a pair of North

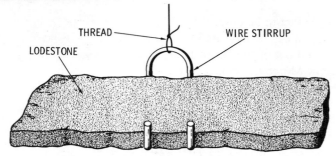

Fig. 1. Lodestone is a magnetic ore.

Fig. 2. A magnetic compass.

poles repel each other, a pair of South poles repel each other, but a North pole attracts a South pole.

Magnetic forces are invisible; however, it is helpful to represent magnetic forces as imaginary lines. For example, we represent the earth's magnetism as shown in Fig. 3. There are several important facts to be observed in this diagram. Since the "North pole" of a compass needle points toward the earth's geographical North pole, we recognize that the earth's geographical North pole has a magnetic South polarity. In other words, the North pole of a compass needle is attracted by magnetic South polarity. This difference between geographic and magnetic poles may seem puzzling at first, but will become clear when we apply the basic law of magnetism to the earth, as illustrated in Fig. 3.

Another important fact shown in Fig. 3 is the location of the earth's magnetic poles with respect to its geographic poles. The earth's magnetic poles are located some distance away from its geographic poles.

10

Still another fact to be observed is that magnetic force lines have a direction, which can be indicated by arrows. Magnetic force lines are always directed out of the North Pole of a magnet and directed into the South pole. Moreover, magnetic force lines are continuous; the lines always form closed paths. Thus, the earth's magnetic force lines in Fig. 3 are continuous through the earth and around the outside of the earth.

The actual source of the earth's magnetism has not been discovered as yet. However, insofar as compass action is concerned, we may imagine that the earth contains a long lodestone along its axis. In turn, this imaginary lodestone will have its South pole near the earth's North geographic pole; the imaginary lodestone will have its North pole near the earth's South geographic pole. It was formerly believed that the earth's magnetism was actually due to the lodestone ore that it

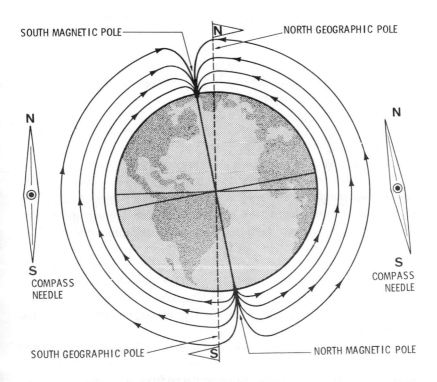

Fig. 3. Earth's magnetic poles.

11

contained. Scientists then proved that this could not be true, and the source of the earth's magnetism is still being sought.

EXPERIMENTS WITH MAGNETS

If we bring the South pole of a magnet near the South pole of a suspended magnet, as shown in Fig. 4, we know that the poles will re-

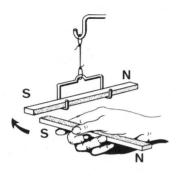

Fig. 4. Showing repulsion between like magnetic poles.

pel each other. It can also be shown that magnetic attractive or repulsive forces vary inversely as the square of the distance between the poles. For example, if we double the distance between a pair of magnetic poles, the force between them is decreased to one-fourth. It can also be shown that if the strength of the magnet in Fig. 4 is doubled (as by holding a pair of similar magnets together with their South poles in the same direction), the force of repulsion is thereby doubled.

The strength of a magnetic field is measured in *gauss*. For example, the strength of the earth's magnetic field is approximately 0.5 gauss; on the other hand, the strongest magnet that has been constructed at present has a strength of about 500,000 gausses. The gauss unit is a measurement of flux density—that is, it is a measure of the number of magnetic force lines that pass through a unit area. *One gauss is defined as one line of force per square centimeter.* In turn, one gauss is equal to 6.452 lines of force per square inch. For example, the strength of the earth's magnetic field is approximately 3.2 lines of force per square inch.

Note that there are 2.54 centimeters in 1 inch, or there is 0.3937 inches in 1 centimeter. Therefore, there are 6.452 square centimeters in 1 square inch, or there is 0.155 square inch in 1 square centimeter. Since 1 gauss is defined as one line of force per square centimeter, it follows that 1 gauss is also equal to 6.452 lines of force per square inch. This book discusses electrical measurements in terms of the English system, insofar as possible. Most electricians prefer the English system. However, various units of measurements have been defined by scientists in terms of the Metric system. If we use suitable conversion factors, we will have no difficulty in converting from the Metric system to the English system.

A unit of magnetic pole strength is measured in terms of force. That is, a *unit magnetic pole* is defined as one which exerts a force of one *dyne* on a similar magnetic pole at a distance of 1 centimeter. If we use a pair of like poles, this will be repulsive force; if we use a pair of unlike poles, it will be an attractive force. There are 444,-800 dynes in 1 pound, or, a dyne is equal to 1/444,800 of a pound. It is not necessary to remember these basic definitions and conversion factors. If you should need them at some future time, it is much more practical to look them up than to try to remember them.

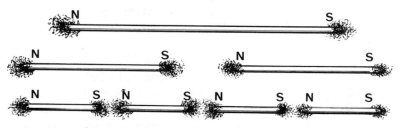

Fig. 5. Showing the effects of breaking a magnet into several parts.

Another important magnet experiment is shown in Fig. 5. If we break a magnetized needle into two parts, each of the parts will become a complete magnet with North and South poles. No matter how many times we break a magnetized needle, we will not obtain a North pole by itself, nor a South pole by itself. This experiment leads us to another basic law of magnetism which states that magnetic poles must always occur in opposite pairs. Many attempts have been made by scientists to find an isolated magnetic pole (magnetic monopole), but all attempts have failed.

We will find that iron and steel are the only substances which can be magnetized to any practical extent. However, there are certain *alloys,*

13

such as *Alnico,* that can be strongly magnetized. Substances such as hard steel and *Alnico* retain their magnetism after they have been magnetized, and are called *permanent magnets.* Since a sewing needle is made from steel, it can be magnetized to form a permanent magnet. On the other hand, soft iron remains magnetized only as long as it is close to or in contact with a permanent magnet. The soft iron loses its magnetism as soon as it is removed from the vicinity of a permanent magnet. Therefore, soft iron forms a *temporary magnet.*

Permanent magnets for experimental work are commonly manufactured from hard steel or magnetic alloys in the form of *horseshoe magnets,* and *bar magnets,* as shown in Fig. 6. The space around the poles

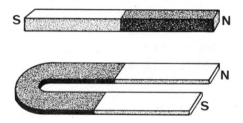

Fig. 6. A bar magnet and a horseshoe magnet.

of a magnet is described as a *magnetic field,* and is represented by magnetic lines of force. The space around a lodestone (Fig. 1), around a compass needle (Fig. 2), around the earth (Fig. 3), and around a permanent magnet (Fig. 4), are examples of magnetic fields. Since a magnetic field is invisible, we can demonstrate its presence only by its force of attraction for iron.

Let us consider the patterns formed by magnetic lines of force in various magnetic fields. One example has been shown in Fig. 3. It can also be easily shown experimentally that when a bar magnet is held under a piece of cardboard, and iron filings are sprinkled on the cardboard, the filing will arrange themselves in curved-line patterns as shown in Fig. 7. The pattern of iron filings that are formed provide a practical basis for our assumption of imaginary lines of force to describe a magnetic field. The total number of magnetic force lines surrounding a magnet, as shown in Fig. 8, is called its total *magnetic flux.*

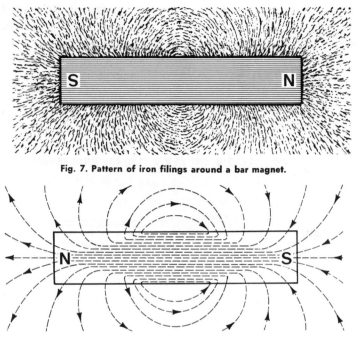

Fig. 7. Pattern of iron filings around a bar magnet.

Fig. 8. Field around a bar magnet represented by lines of force.

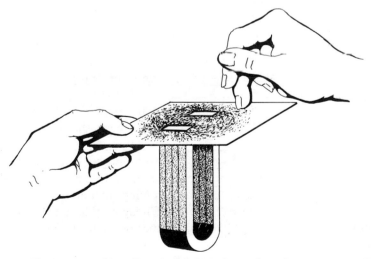

Fig. 9. Pattern of iron filings in the space above a horseshoe magnet.

15

A similar experiment with a horseshoe magnet is shown in Fig. 9. The iron filings arrange themselves in curved lines that suggest the imaginary lines of force that we use to describe a magnetic field. Note that the magnetic field is strongest at the poles of the magnet in Fig. 7. Since the field strength falls off as the square of the distance from a pole, a magnet exerts practically no force on a piece of iron at an appreciable distance. A magnet exerts its greatest force on a piece of iron when in direct contact.

FORMATION OF PERMANENT MAGNETS

Another important and practical experiment is the magnetization of steel to form a permanent magnet. For example, if we wish to magnetize a steel needle, we may use any of the following methods:

1. The needle can be stroked with one pole of a permanent magnet. The needle can be stroked several times to increase its magnetic strength, but each stroke must be made in the same direction.
2. If the needle is held in a magnetic field (such as between the poles of a horseshoe magnet) and the needle is tapped sharply, it will become magnetized.
3. We can heat a needle to dull red heat and then quickly cool the needle with cold water while holding it in a magnetic field, and the needle will become magnetized.

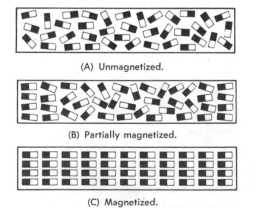

(A) Unmagnetized.

(B) Partially magnetized.

(C) Magnetized.

Fig. 10. Representation of molecular magnets in a steel bar.

The formation of permanent magnets is explained in terms of molecular magnets. Each molecule in a steel bar is regarded as a tiny permanent magnet. As shown in Fig. 10, the poles of these molecular magnets are distributed at random in an unmagnetized steel bar. Therefore, the fields of the molecular magnets cancel out on the average, and the steel bar does not act as a magnet. On the other hand, when we stroke an unmagnetized steel bar with the pole of a permanent magnet, some of the molecular magnets respond by lining up end-to-end. In turn, the lined-up molecular magnets have a combined field that makes the steel bar a magnet. If the steel bar is stroked a number of times, more of the molecular magnets are lined up end-to-end, and a stronger permanent magnet is formed as shown in Fig. 11A.

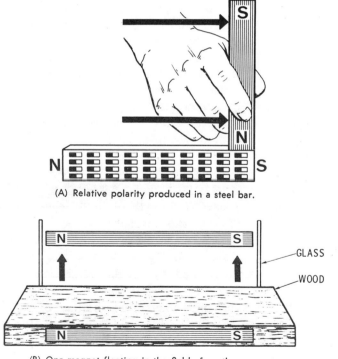

(A) Relative polarity produced in a steel bar.

(B) One magnet floating in the field of another magnet.

Fig. 11. Magnet characteristics.

Steel is much harder than iron; in turn, it is more difficult to line up the molecular magnets in a steel bar than in a soft-iron bar. Therefore, to make a strong permanent magnet from a steel bar, we must stroke the bar many times with a strong permanent magnet. A soft-iron bar becomes fully magnetized as soon as it is touched by a permanent magnet, but will return to its unmagnetized state as soon as it is removed from the field of a permanent magnet. Once the molecular magnets have been lined up in a hard steel bar, however, they will retain their positions and provide a permanent magnet.

Although there are a very large number of molecules in an iron or steel bar, there are not an infinite number of molecules to be lined up. Therefore, there is a limit to which the bar can be magnetized, no matter how strong a field we use. When all the molecular magnets are aligned in the same direction, the bar cannot be magnetized further, and the iron or steel is said to be *magnetically saturated*. The ability of a magnetic substance to retain its magnetism after the magnetizing force has been removed is called its *retentivity*. Thus, retentivity is very large in hard steel, and almost absent in soft iron. Magnetic alloys such as *Alnico V* have a very high retentivity and are widely used in modern electrical and electronic equipment. The *Alnico* alloys contain iron, aluminum, nickel, copper, and cobalt in various proportions depending on the requirements.

A permanent magnet that weighs 1½ pounds may have a strength of 900 gauss, and will lift approximately 50 pounds of iron. This type of magnet is constructed in a horseshoe form, and is less than 3-inches long. A 5-pound magnet may have a strength of 2000 gauss, and will lift approximately 100 pounds of iron. A 16-pound magnet 5½-inches long may have a strength of 4800 gauss and will lift about 250 pounds of iron. Bar magnets can be magnetized with sufficient strength that one of the magnets will float in the field of the other magnet, as shown in Fig. 11B. A similar demonstration of magnetic forces is provided by circular ceramic magnets. Each circular magnet is about 2½-inches in diameter and has a hole in the center that is 1-inch in diameter. One surface of the disc is a North pole, and the opposite surface is a South pole. When placed on a nonmagnetic restraining pole, with like poles adjacent, the circular magnets float in the air, being held in suspension by repelling magnetic forces.

AIDING AND OPPOSING MAGNETIC FIELDS

An experiment that demonstrates the repulsion of like magnetic poles was illustrated in Fig. 4. The question is often asked how magnetic force lines act in aiding or opposing magnetic fields. Fig. 12

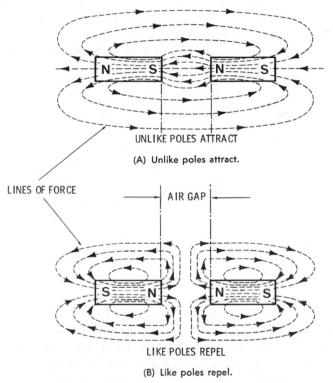

UNLIKE POLES ATTRACT

(A) Unlike poles attract.

LINES OF FORCE

AIR GAP

LIKE POLES REPEL

(B) Like poles repel.

Fig. 12. Lines of force between unlike and like poles.

shows the answer to this question. Note that when unlike poles are brought near each other, the lines of force in the air gap are in the *same* direction. Therefore, these are aiding fields and the lines concentrate between the unlike poles. It is a basic law of magnetism that lines of force tend to shorten as much as possible—lines of force have been compared to rubber bands in this respect. In turn, a force of attraction is exerted between the unlike poles in Fig. 12A.

On the other hand, a pair of like poles have been brought near each other in Fig. 12B. The lines of magnetic force are directed in opposition, and the lines from one pole oppose the lines from the other pole. In turn, none of the lines from one magnet enter the other magnet, and a force of repulsion is exerted between the magnets. We observe that the magnetic fields in Fig. 12 are changed in shape, or are *distorted* with respect to the field shown in Fig. 8. If the magnets in Fig. 12 are brought more closely together, the fields become more distorted. Hence, we recognize that forces of attraction or repulsion between magnets are produced by distortion of their magnetic fields.

ELECTROMAGNETISM

Electromagnetism is the production of magnetism by an electric current. An electric current is a flow of electrons; we can compare the flow of electrons in a wire with the flow of water in a pipe. Today, we can read about electronics and electrons in the newspapers, magazines, and many school books. However, the practical electrician needs to know more about electrons than a mechanic or machinist. Therefore, let us see how electric current flows in a wire.

An atom is the smallest particle of any substance; thus, the smallest particle of copper is a copper atom. We often hear about *splitting the atom*. If a copper atom is split or broken down into smaller particles, we find that it is built up from extremely small particles of electricity. In other words, all substances such as copper, iron, and wood have the *same* building blocks, and these building blocks are particles of electricity. Copper and wood are different substances simply because these particles of electricity are arranged differently in their atoms. An atom can be compared with our solar system in which the planets revolve in orbits around the sun. For example, a copper atom has a *nucleus* which consists of positive particles of electricity; electrons (negative particles of electricity) revolve in orbits around the nucleus.

Fig. 13 shows three atoms in a metal wire. An electron in one atom can be transferred to the next atom under suitable conditions, and this movement of electrons from one end of the wire to the other end is called an electric current. Electric current is *electron flow*. To make electrons flow in a wire, an *electrical pressure* must be applied to the ends of the wire. This electrical pressure is a force named *electromotive*

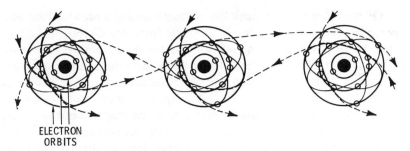

ELECTRON
ORBITS

Fig. 13. Atoms in a metal wire.

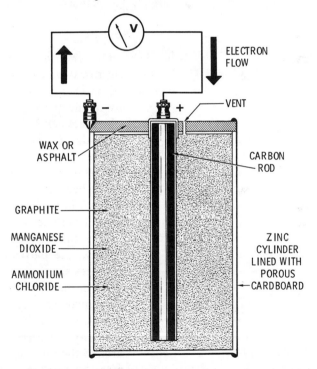

ELECTRON
FLOW

VENT

WAX OR
ASPHALT

CARBON
ROD

GRAPHITE

MANGANESE
DIOXIDE

ZINC
CYLINDER
LINED WITH
POROUS
CARDBOARD

AMMONIUM
CHLORIDE

Fig. 14. A dry cell produces electromotive force by chemical action

force, or *voltage.* For example, an ordinary dry cell is a source of
electromotive force. A dry cell produces electromotive force by chemical action. If we connect a voltmeter across a dry cell, as shown in

21

Fig. 14, electrons flow through the voltmeter, which indicates the voltage of the cell.

We measure electromotive force (emf) in *volts*. If a dry cell is in good condition, it will have an emf of about 1.5 volts. Note that a dry cell acts as a *charge separator*. In other words, the chemical action in the cell takes electrons away from the carbon rod and adds electrons to the zinc cylinder. Therefore, there is an electron pressure or emf at the zinc cylinder. When a voltmeter is connected across a dry cell (Fig. 14), this electron pressure forces electrons to flow in the connecting wire as shown in Fig. 13. We observe that electrons flow from the negative terminal of the dry cell, around the wire *circuit,* and back to the positive terminal of the dry cell.

Next, we will find that an electric current produces a magnetic field. For example, if a compass needle is brought near a current-carrying wire, the compass needle turns as shown in Fig. 15. Since a compass

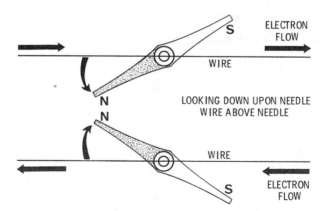

Fig. 15. A compass needle is deflected in the vicinity of a current-carrying wire.

needle is acted upon by a magnetic field, this experiment shows that the electric current is producing a magnetic field. This is the principle of electromagnetism. The magnetic lines of force surrounding a current-carrying wire can be demonstrated as shown in Fig. 16. When iron filings are sprinkled over the cardboard, the filings arrange themselves in circles around the wire.

Although a current-carrying wire acts as a magnet, it is a form of *temporary magnet*. The magnetic field is not present until the wire is connected to the dry cell. Magnetic force lines are produced only while there is current in the wire. As soon as the circuit is opened (disconnected from the dry cell), the magnetic force lines disappear. Let us

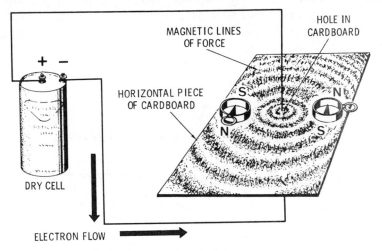

Fig. 16. Demonstration of electromagnetism.

observe the polarities of the compass needles in Fig. 16. The magnetic lines of force are directed clockwise, looking down upon the cardboard. This experiment leads us to a basic rule of electricity called the *left-hand rule*. Fig. 17 illustrates the left-hand rule; if a conductor is grasped with the left hand, with your thumb pointing in the direction of electron flow, then your fingers will point in the direction of the magnetic lines of force.

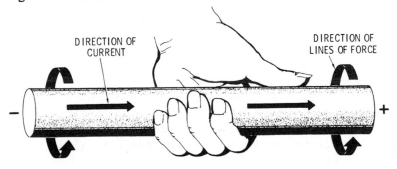

Fig. 17. Left-hand rule used to determine direction of magnetic force lines around a current-carrying conductor.

Experiments show that the magnetic field around the wire in **Fig. 16** is weak, and we will now ask how the strength of an electromagnetic field can be increased. The magnetic field around a straight wire is comparatively weak because it is produced over a large volume of space. To reduce the space occupied by the magnetic field, a straight wire can be bent in the form of a loop, as shown in Fig. 18. Now, the

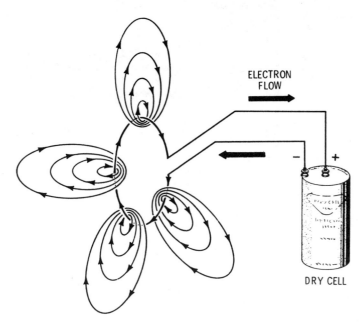

ELECTRON
FLOW

DRY CELL

Fig. 18. The magnetic field is concentrated by forming a conductor into a loop.

magnetic flux lines are concentrated in the area enclosed by the loop. Therefore, the magnetic field strength is comparatively great inside the loop. This is an elementary form of *electromagnet.*

Next, to make an electromagnet with a much stronger magnetic field, we can wind a straight wire in the form of a helix with a number of turns, as shown in Fig. 19. Since the field of one loop adds to the field of the next loop, the total field strength of the electromagnet is much greater than if a single turn were used. Note that if we use the same wire shown in Fig. 16 to form the electromagnet in Fig. 19, the current is the same in both circuits. That is, we have not changed the

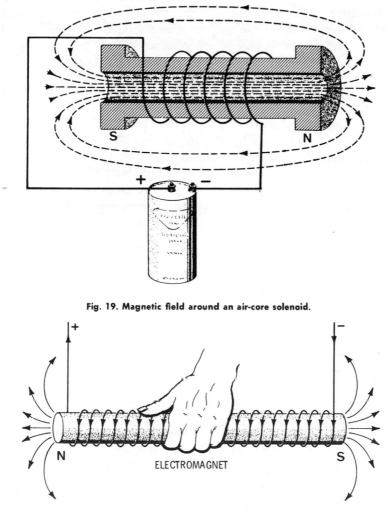

Fig. 19. Magnetic field around an air-core solenoid.

Fig. 20. Showing method of finding the magnetic polarity of a coil by means of the left-hand rule.

amount of current—we have merely concentrated the magnetic flux by winding the wire into a spiral. Electricians often call an electromagnet of this type a *solenoid*.

The name "solenoid" is applied to electromagnets that have an *air core*. For example, we might wind the coil in Fig. 19 on a wooden spool. Since wood is not a magnetic substance, the electromagnet is essentially an air-core magnet. Note the polarity of the magnetic field in Fig. 19 with respect to the direction of current flow. The left-hand rule applies to electromagnets, just as to straight wires. Thus, if we grasp an electromagnet as shown in Fig. 20, with the fingers of the left hand in the direction of electron flow, then the thumb will point to the North pole of the electromagnet.

Since iron is a magnetic substance, the strength of an electromagnet can be greatly increased by placing an iron core inside a solenoid. For example, if we place a soft-iron bar inside the wooden spool in Fig. 19, we will find that the magnetic field strength becomes much greater. Let us see why this is so (with reference to Fig. 10); the molecular magnets in the soft-iron bar, or *core*, are originally oriented in random directions. However, under the influence of the flux lines inside the electromagnet, these molecular magnets line up in the same direction. Therefore, the magnetic field of the molecular magnets is added to the magnetic field produced by the electric current, and the strength of the electromagnet is greatly increased.

Note that we have not changed the amount of current in the wire by placing an iron core in the solenoid. The magnetic field produced by the electric current remains unchanged. However, the electromagnet has a much greater field strength when an iron core is used because we then have two sources of magnetic field which add up to produce the total field strength. If we open the circuit of an iron-core electromagnet, both sources of magnetic field disappear. In other words, both solenoids and iron-core electromagnets are temporary magnets.

VOLTS, AMPERES, AND OHMS

To fully understand circuits such as shown in Fig. 19, we must recognize another basic law of electricity called *Ohm's law*. This is a simple law that states the relation between *voltage, current,* and *resistance*. We have become familiar with voltage, and we know that voltage (emf) is an electrical pressure. We also know that an electromotive force causes electric charges (electrons) to move through a wire. Electron flow is called an electric *current*, and a wire opposes (resists) electron flow. This opposition is called electrical *resistance*. We measure electromotive force (voltage) in *volts;* we measure electric current in *amperes;* we measure resistance in *ohms*.

Fig. 21. Appearance of a voltmeter.

Fig. 22. Appearance of an ammeter.

Fig. 23. View of an ohmmeter.

Fig. 14 showed how the voltage of a dry cell is measured with a voltmeter. A commercial voltmeter is illustrated in Fig. 21. Current is measured with an *ammeter,* as illustrated in Fig. 22. Resistance is measured with an ohmmeter, as illustrated in Fig. 23. At this time, we are interested in the relation between voltage, current, and resistance. Ohm's law states that the current in a circuit is directly proportional to the applied voltage, and inversely porportional to the circuit resistance. Thus, we write Ohm's law as follows:

$$\text{Electric current} = \frac{\text{electromotive force}}{\text{resistance}}$$

For example, Ohm's law states that if a wire has 1 ohm of resistance, an emf of 1 volt applied across the ends of the wire will cause 1 ampere of current to flow through the wire. Or, if we apply 2 volts across

1 ohm of resistance, 2 amperes of current will flow. Again, if we apply 1 volt across 2 ohms of resistance, ½ ampere of current will flow. We generally write Ohm's law with letters, as follows:

$$I = \frac{E}{R}$$

where,

I represents current in amperes,
E represents emf in volts,
R represents resistance in ohms.

Note that electrons (electric charges) flow in an electric circuit. On the other hand, voltage does not flow; resistance does not flow. Electric current is defined as the *rate* of charge flow. A current of 1 ampere consists of 6.24×10^{18} electrons flowing past a point in one second. One *coulomb* is defined as 6.24×10^{18} electrons (6,240,-000,000,000,000,000 electrons). Therefore, a current of 1 ampere denotes the passage of 1 coulomb past a point in 1 second. Although we often speak of "current flow", we really refer to charge flow, because current denotes the *rate of charge flow*. If electrons flow at the rate of 3.12×10^{18} electrons per second, the current value is ½ ampere.

ELECTRIC AND MAGNETIC CIRCUITS

To measure the current in a circuit, we connect an ammeter into the circuit as shown in Fig. 24. It is a basic law of electricity that the current is the same at any point in the circuit. Therefore, the ammeter indicates the amount of current that flows in *each turn* of the electromagnet. Each turn produces a certain amount of magnetism in the core. As we would expect, the amount of magnetism that is produced by each turn of wire depends on the amount of current in the wire. Therefore, we describe an electromagnet in terms of *ampere-turns*. If the ammeter in Fig. 24 indicates a current of 1 ampere, each turn on the coil represents 1 ampere-turn.

29

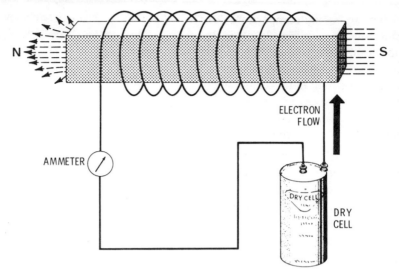

Fig. 24. Measurement of current in the circuit.

The total number of ampere-turns on a coil is equal to the number of amperes times the number of turns. For example, with a current of 1 ampere in Fig. 24, there will be a total of 9 ampere-turns. Next, let us suppose that we increase the current in Fig. 24 to 2 amperes. Now, 2 amperes of current flows in each turn of the coil, and we have a total of 18 ampere-turns. Since the strength of the electromagnet is proportional to the number of its ampere-turns, *we double the strength of the electromagnet by doubling the amount of current.*

Since each ampere-turn produces a certain amount of magnetism, the ampere-turn is taken as the unit of *magnetomotive force.* Magnetomotive force is measured in ampere-turns. We often compare magnetomotive force with electromotive force. In other words, electromotive force produces current in a wire, and magnetomotive force produces magnetic flux lines in a core. We will find that any core, such as air or iron, opposes the production of magnetic flux lines. This opposition is called the reluctance of the core. We often compare reluctance with resistance. In other words, production of magnetic flux lines is opposed by reluctance, and production of electric current is opposed by resistance. Therefore, we can also compare magnetic flux lines with electric current.

The foregoing comparisons lead us to the idea of a *magnetic circuit,* as shown in Fig. 25. This diagram shows both an electric circuit and a magnetic circuit. The electric circuit consists of the 1-volt battery and the 1-ohm coil through which 1 ampere of current flows. The magnetic circuit in this example consists of a circular iron core. Reluctance is measured in *rels*. This core has a reluctance of 1 rel. A basic law of electromagnetism states that 1 ampere-turn produces 1 line of magnetic flux in a reluctance of 1 rel. Therefore, we write a law for magnetic circuits that is much the same as Ohm's law for electric circuits:

$$\text{magnetic flux} = \frac{\text{magnetomotive force}}{\text{reluctance}}$$

or,

$$\text{flux lines} = \frac{\text{ampere-turns}}{\text{rels}}$$

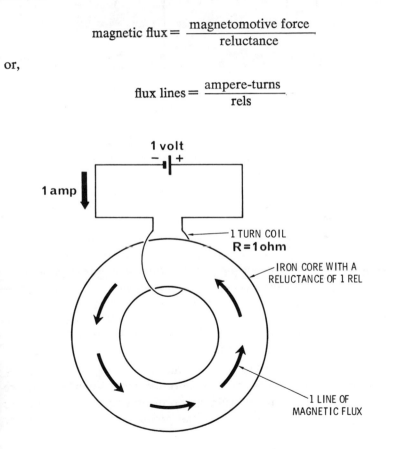

Fig. 25. The basic magnetic circuit.

31

Next, suppose that we increase the current in Fig. 25 to 2 amperes. Then, 2 lines of magnetic flux will be produced in the iron core. Another example is shown in Fig. 26, where 5 amperes of current flows

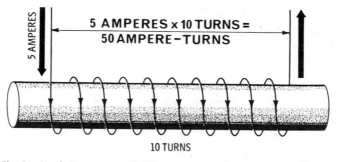

Fig. 26. An electromagnet with 50 ampere-turns of magnetomotive force.

through 10 turns, providing a magnetomotive force of 50 ampere-turns. The number of flux lines that will be produced in Fig. 26 depends on the reluctance of the magnetic circuit. Note that a continuous iron magnetic circuit is provided for the flux lines in Fig. 25; the magnetic circuit in Fig. 26 is more complicated, however, because part of the magnetic circuit is iron and the other part is air. From the previous discussion of electromagnets, we would expect that air has a much greater reluctance than iron.

A volume of air that is 1 inch square and 3.19 inches long has a reluctance of 1 rel. This is 1500 times the reluctance of a typical iron core of the same size. If we place a closed iron core in a solenoid, the number of flux lines will increase 1500 times in a typical experiment. However, the following are several facts that we must keep in mind when working with iron cores:

1. Different types of iron have different amounts of reluctance.
2. The reluctance of iron changes as the magnetomotive force is changed.
3. Since air has a much greater reluctance than iron, a magnetic circuit with a large air gap acts practically the same as an air core.

For example, cast iron has about 6 times as much reluctance as annealed sheet steel. To show the change in the reluctance of iron

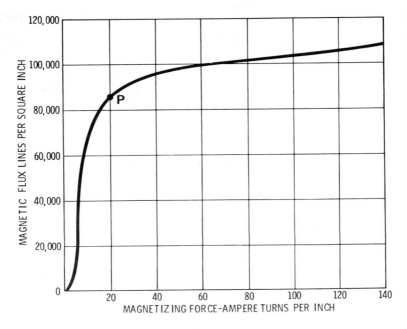

Fig. 27. Magnetizing force versus flux density for annealed sheet steel.

when the magnetomotive force is changed, we use charts such as shown in Fig. 27. The chart shows a magnetization curve. Note that the scales on the chart are marked off in terms of *magnetizing force per inch*, and *magnetic flux lines per square inch*. Magnetizing force is equal to magnetomotive force per inch of core length. For example, let us consider the iron core shown in Fig. 28. The length of this magnetic circuit is 16 inches. If we apply a magnetomotive force of 320 ampere-turns to this core, we will have a magnetizing force of 20 ampere-turns per inch.

Let us find how many flux lines per square inch will be produced in the core (Fig. 28) when a magnetizing force of 20 ampere-turns per inch is applied. At point P on the magnetization curve in Fig. 27, we see that there will be about 83,000 lines per square inch produced in the core. Now, if the core in Fig. 28 has a cross-sectional area of 1 square inch, there will be 83,000 lines of magnetic flux in the core. Or, if the core has a cross-sectional area of 2 square inches, there will be 166,000 lines of magnetic flux in the core. Again, if the core

33

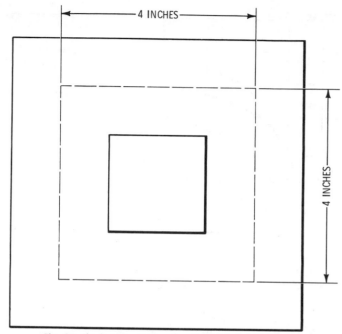

Fig. 28. The length of this magnetic circuit is 16 inches.

has a cross-sectional area of ½ square inch, there will be 41,500 lines of magnetic flux in the core.

The number of magnetic flux lines per square inch is generally called the *flux density* in the core; flux density is represented by the letter B, and magnetizing force is represented by the letter H. Thus, a magnetization curve such as shown in Fig. 27 is usually called a *B-H curve*. Note that the flux density in this example increases rapidly from 0 to 20 ampere-turns per inch. At higher values of magnetizing force, the B-H curve flattens off. This flattened-off portion of the curve is called the *saturation interval*. The curve will finally become horizontal, and the iron core then has the same reluctance as air. When an iron core is completely saturated, we can remove the iron core from the electromagnet, and its magnetic field strength will remain the same.

Fig. 29 shows some comparative B-H curves. We observe that an electromagnet with an annealed sheet-steel core will have a much stronger magnetic field than if cast iron is used for a core. Note also

Fig. 29. Comparative B-H curves.

that if we wish to make a very strong electromagnet, it is better to use a large cross-sectional area in the core instead of a large number of ampere-turns per inch. In other words, iron and steel starts to saturate when more than 20 ampere-turns per inch are used, and core saturation corresponds to wasted electric current. Therefore, efficient operation requires that we use no more than 20 ampere-turns per inch, and then make the cross-sectional area of the core as large as required to obtain the desired number of magnetic flux lines.

UNDERSTANDING ELECTRIC CIRCUITS

Any circuit must contain a voltage source to be of practical use. Source voltages may be very high, moderate, or very low. A dry cell is a familiar 1.5 volt source, as was shown in Fig. 14. When a higher voltage is required, cells can be connected in *series* to form a *battery*, as shown in Fig. 30. Note that the negative terminal of one cell is con-

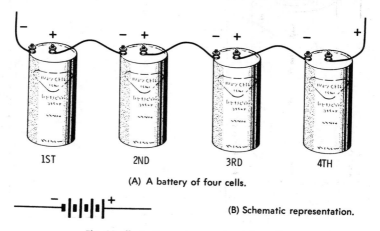

1ST   2ND   3RD   4TH

(A) A battery of four cells.

(B) Schematic representation.

Fig. 30. Illustrating series connected dry cells.

nected to the positive terminal of the next cell. This series connection causes the cell voltages to be added. Since there are four cells in the example, the battery voltage will be approximately 6 volts. New dry cells usually have an emf of slightly more than 1.5 volts. As a cell ages, its emf decreases.

Current does not flow in a circuit unless it is *closed*. Fig. 31 shows the difference between a closed circuit and an open circuit. A flashlight bulb might draw 0.25 ampere from a 1.5-volt source. Let us apply Ohm's law to find the resistance of the bulb. It is easy to see that there are three possible arrangements of Ohm's law, as shown in Fig. 32. Since we are interested in finding the resistance of the bulb, we will use the second arrangement of Ohm's law. Accordingly, the resistance of the bulb is 1.5/0.25, or 6 ohms. Next, let us observe how the other two arrangements of Ohm's law are used.

In case we know the applied voltage (1.5 volts) and the resistance of the bulb (6 ohms), we will use the first arrangement of Ohm's law

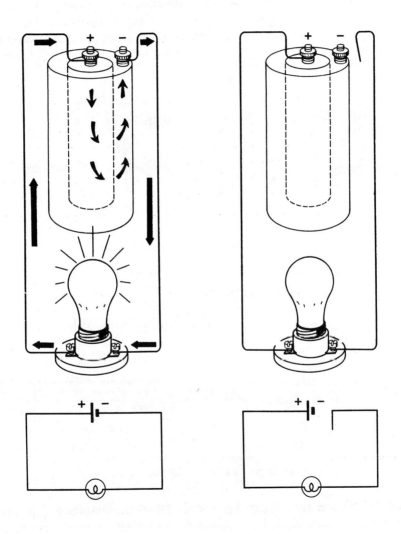

(A) Closed circuit. (B) Open circuit.

Fig. 31. Simple electric circuit.

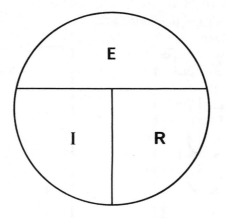

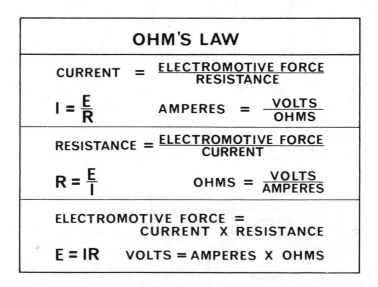

Fig. 32. Ohm's law in diagram form.

to find the current in the circuit. Thus, the current is equal to 1.5/6, or 0.25 ampere. On the other hand, in case we know the current through the bulb (0.25 ampere) and the resistance of the bulb (6 ohms), we will use the third arrangement of Ohm's law to find the applied voltage. Thus, the applied voltage is equal to 0.25 × 6, or 1.5 volts. Fig. 32 shows Ohm's law in diagram form. To find one of the quantities, first cover that quantity with a finger. The location of the other two letters in the circle will then show whether to divide or multiply. For example to find I, we cover I and observe that E is to be divided by R. Again, to find E, we cover E and observe that I is to be multiplied by R.

A circuit can be opened by disconnecting a wire as shown in Fig. 31B. To conveniently open and close a circuit, a switch is used as shown in Fig. 33. This type of switch is called a knife switch. Note

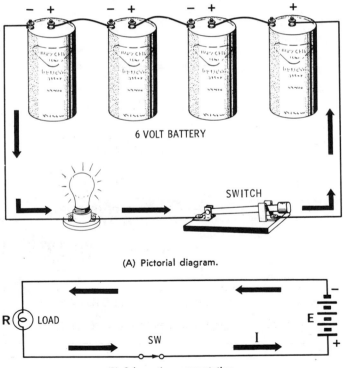

(A) Pictorial diagram.

(B) Schematic representation.

Fig. 33. Illustrating a simple electric circuit.

39

that the lamp is called a *load resistance,* or simply a *load.* An electrical load in a circuit changes electricity into light and heat, as in this example; in other circuits, the load may change electricity into mechani-

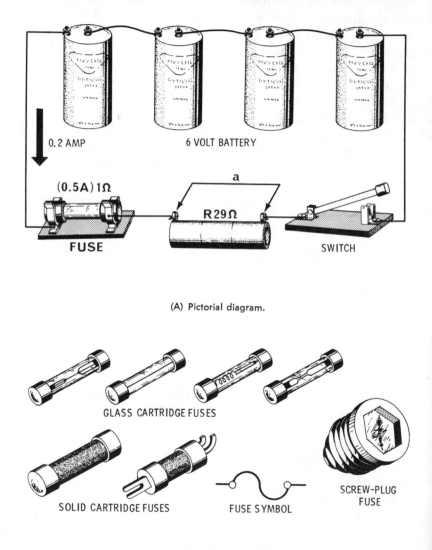

(A) Pictorial diagram.

(B) Typical fuses.

Fig. 34. A fused circuit.

cal power or some other form of power. Let us consider the action of the *fuse* shown in Fig. 34. A fuse is a type of automatic safety switch; the fuse "blows" and opens the circuit in case the load (R) should become short-circuited and draw excessive current from the battery.

Fuses are made from thin strips or wire of aluminum or other metal. The resistance of a fuse is comparatively low, but because of its small cross-section, the fuse heats up and melts if a certain amount of current flows through it. For example, the fuse shown in Fig. 34A has a resistance of 1 ohm. The load R has a resistance of 29 ohms, making a total of 30 ohms of circuit resistance. Since 6 volts are applied, 0.2 ampere will flow in accordance with Ohm's law. This fuse is made so that it will blow at a current of 0.5 ampere. Therefore, the fuse does not blow as long as R is not short-circuited.

We could short-circuit R by placing a screwdriver across the terminals. In such case, the total circuit resistance would be reduced to 1 ohm, and 6 amperes of current would flow. Therefore, the fuse would immediately blow out and automatically open the circuit. Thereby, the battery is protected from damage due to excessive current drain. Note that after the fuse blows, the voltage between its terminals will be 6 volts. In normal circuit operation, the voltage across the load-resistor terminals (a) is 5.8 volts, in accordance with Ohm's law. The voltage across the fuse terminals is 0.2 volt, in accordance with Ohm's law. We call the voltage between the load-resistor terminals the *voltage drop* across the resistor; similarly, we call the voltage between the fuse terminals the *voltage drop* across the fuse.

When we speak of a voltage drop, we simply mean that this amount of voltage would be measured by a voltmeter connected across the terminals of resistance. It follows from the foregoing example that *Ohm's law applies to any part of a circuit, as well as to the complete circuit.* In other words, insofar as the load resistor is concerned in Fig. 34A, 0.2 ampere is flowing through 29 ohms; therefore, the IR drop across the load resistor is 5.8 volts. Insofar as the fuse is concerned, 0.2 ampere is flowing through 1 ohm; therefore, the IR drop across the fuse is 0.2 volt.

We have noted previously that both electromotive force (emf) and voltage are measured in volts. Electromotive force is often called a voltage. Nevertheless, there is a basic distinction between an emf and a voltage. For example, a dry cell is a chemical *source* of voltage, and we speak of the *emf produced by the source* of electricity. This emf is measured in volts. If a resistor is connected across a source of electricity, the current produces a voltage drop across the resistance in accordance with Ohm's law. This voltage drop is not called an emf, but is called a voltage, and is measured in volts. In other words, when we speak in strict terms, we speak of the emf of a voltage source, but we speak of the voltage produced across a load resistor.

KIRCHHOFF'S VOLTAGE LAW

The foregoing example also illustrates another basic law of electric circuits called Kirchhoff's voltage law. This law states that *the sum of the voltage drops around a circuit is equal to the source voltage.* Note that the sum of the voltage drops across the load resistor and the fuse is equal to 6 volts (5.8 + .2), and that the source voltage (battery voltage) is also equal to 6 volts. We recognize that Kirchhoff's voltage law is simply a summary of Ohm's law as applied to all the resistances in a circuit. Although we do not need to use Kirchhoff's voltage law in describing the action of simple circuits, this law will be found very useful in solving complicated circuits.

ELECTRICAL POWER

There are many forms of power. For example, an electric motor produces a certain amount of mechanical power, usually measured in horsepower. An electric heater produces heat (thermal) power. An electric light bulb produces both heat power and light power (usually measured in candlepower). Electrical power is measured in *watts;* electrical power is equal to volts times amperes. Thus, we write:

$$\text{watts} = \text{volts} \times \text{amperes}$$

or,

$$P = EI$$

With reference to Fig. 34A, the battery supplies $6 \times 0.2 = 1.2$ watts to the circuit; the load resistor R takes $5.8 \times 0.2 = 1.16$ watts; the fuse takes $0.2 \times 0.2 = 0.04$ watt. Note that the power taken by both re-sistances is equal to $1.16 + 0.04 = 1.2$ watts. In other words, the power supplied by the battery is exactly equal to the power taken by the circuit resistance. This fact leads us to another basic law called the *law of conservation of energy*. This law states: *Energy cannot be created nor destroyed, but only changed into some other form of energy.* This is the same as saying that power can only be changed into some other form of power, because energy is equal to power multiplied by time.

In the example of Fig. 34, electrical energy is changed into heat energy (or electrical power is changed into heat power) by the load resistor and the fuse. Since the load resistor takes 1.16 watts of electrical power, it produces 1.16 watts of heat power. With reference to Fig. 33, light is measured in candlepower. An ordinary electric-light bulb produces approximately 1 candlepower for each watt of electrical power. For example, a 60-watt lamp normally takes 60 watts of electrical power and produces about 60 candlepower of light.

Since power is equal to IE, and $I = E/R$, we can write $P = E^2/R$. Since power is equal to IE, and $E = IR$, we can write $P = I^2R$. Thus, by substitution from Ohm's law into the basic power law, and by rearranging these equations, we obtain the 12 important electrical formulas shown in Fig. 35. In summary, these formulas state:

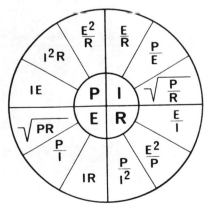

Fig. 35. Summary of basic formulas.

$$I = \frac{E}{R} = \frac{P}{E} = \sqrt{\frac{P}{R}}$$

$$E = IR = \frac{P}{I} = \sqrt{PR}$$

$$R = \frac{E}{I} = \frac{P}{I^2} = \frac{E^2}{P}$$

$$P = IE = I^2R = \frac{E^2}{R}$$

These formulas are important because we may be working with a circuit in which the voltage and resistance are known, and we need to find the amount of current flow. Again, we may know the voltage and power in the circuit, and need to find the amount of current flow. Or, we may know the resistance and power in the circuit, and need

Courtesy Sargent-Welch Scientific Co.

Fig. 36. View of a wattmeter.

to find the amount of current flow. Similarly, we can find the voltage of a circuit if we know the current and resistance, or the current and the power, or the resistance and the power. We can find the resistance in a circuit if we know the voltage and current, or the current and power, or the voltage and power. We can find the power in a circuit if we know the voltage and current, or the current and resistance, or the voltage and resistance.

Power is measured with a wattmeter, as illustrated in Fig. 36. Next, let us consider the measurement of electrical energy. We measure electrical energy in watt-seconds, watt-hours, or kilowatt-hours. A watt-second is equal to the electrical energy produced by 1 watt in 1 second. In turn, a watt-hour is equal to the electrical energy produced

Fig. 37. Watt-hour meter.

by 1 watt during 1 hour; therefore, a watt-hour is equal to 3600 watt-seconds. A kilowatt-hour is equal to 1000 watt-hours. An instrument used to measure electrical energy is called a watt-hour meter, as illustrated in Fig. 37. We are quite familiar with watt-hour meters, because they are installed by public utility companies in every home, office, and shop.

QUICK-CHECK INSTRUMENTS FOR TROUBLESHOOTING

Electrical troubleshooting requires various tests. However, practical electricians generally make quick-checks with continuity testers and voltage indicators. Shop work often requires the use of voltmeters, ammeters, ohmmeters, and wattmeters. A typical continuity tester is illustrated in Fig. 38. It consists of a dry cell and a doorbell, with a pair of test leads. It is usually preferred when connections are to be checked in a wiring system, or when a broken place in a wire is to be located. If the test leads are applied across a closed circuit, the bell rings. Note that if a poor connection is being checked, the bell may or may not ring, depending on the resistance of the bad connection.

A typical voltage indicator is illustrated in Fig. 39. It consists of a neon bulb with a built-in dropping resistor and a pair of

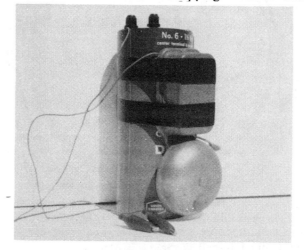

Fig. 38. An electrician's continuity tester.

test leads. This type of tester is usually preferred when "hot" and "cold" circuits are being checked in a wiring system. If the test leads are applied across a circuit in which 70 volts or more are present, the neon bulb will glow. The bulb will glow much brighter across a 240-volt circuit than across a 120-volt circuit. However, the brightness of the glow cannot be used to estimate the voltage value accurately. A neon tester is essentially a quick-checker. However, it is a very practical tester in general trouble-shooting procedures.

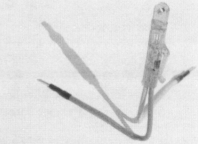

Fig. 39. An electrician's voltage indicator.

SUMMARY

A lodestone is a natural magnet consisting chiefly of a magnetic oxide of iron called magnetite. All magnets have a North and South pole. Magnetic lines of force are invisible but arc always continuous, always forming a closed path. Like poles repel, unlike poles attract.

Iron and steel are the only material which can be magnetized to any practical extent. Certain alloys, such as *Alnico,* can be strongly magnetized and are called permanent magnets. Permanent magnets are generally in the form of a horseshoe or bar. The space around the poles of a magnet is described as a magnetic field, and is represented by magnetic lines of force. The total number of magnetic force lines surrounding a magnet is called its total magnetic flux.

Electromagnetism is a production of magnetism by an electric current. Electric current is a flow of electrons which is compared to the flow of water in a pipe. Electric current is electron flow; and to make electrons flow, pressure must be applied to the end of the wire. This pressure applied is called the electromotive force or voltage. Electromotive force (emf) is measured in volts.

To fully understand electromagnetism or the basic laws of electricity, we must use Ohm's law. This is a law that states the relationship between voltage, current, and resistance. Voltage is the electrical pressure that causes electrons to move through a wire. Electron flow is called an electric current. Current is the movement of electrons through a conductor. A wire (or conductor) opposes the electron flow because of resistance.

TEST QUESTIONS

1. What is a lodestone?
2. Do unlike magnetic poles attract or repel each other?
3. Why does a compass needle point in a North-and-South direction?
4. How does a permanent magnet differ from a temporary magnet?
5. Can a North pole exist without a South pole?
6. Are magnetic lines of force directed into or out of the North pole of a magnet?
7. What is the definition of electromagnetism?
8. How do metals conduct electricity?
9. Is a dry cell a source of magnetism or of electricity?
10. In what way is electromotive force, or voltage, an electrical pressure?
11. Will a voltmeter measure the voltage or the current of a dry cell?
12. Why is a compass needle deflected in the vicinity of a current-carrying wire?
13. How is an electromagnet constructed?
14. Can you explain why a soft-iron core increases the strength of an electromagnet?
15. What is the law named that relates voltage, current, and resistance?
16. Is a magnetic circuit the same thing as an electric circuit?
17. How is an ampere-turn defined?
18. Does an ammeter measure current or voltage?
19. Can you state a law for magnetic circuits that is similar to Ohm's law?
20. Why are dry cells connected in series?
21. What is the meaning of a closed circuit? An open circuit?
22. In what way can a fuse be compared with a switch?

CHAPTER 2

Conductors
and Insulators

A conductor is a substance that carries electric current. An insulator
is a substance that does not carry electric current. Since no conductor
is perfect, and because any conductor has at least a small amount
of resistance, it is better to define a conductor as a substance with a
very low resistance. We will also find that no insulator is perfect, and
because no insulator has an infinite resistance, it is better to define an
insulator as a substance with a very high resistance. Therefore, conductors, resistors, and insulators are basically resistive substances. However, they are classified into different groups, because a practical conductor has extremely low resistance, a load resistor has a moderate
resistance, and a good insulator has extremely high resistance.

CLASSES OF CONDUCTORS

The substances listed in Table 1 have different conductivities. The
best conductors are listed in the first column, and in the order that they
appear. For example, silver is the best conductor, lead has less conduc-

tivity, carbon has still less conductivity, moist earth is a poorer conductor than carbon, and slate has such a high resistance that it is called an insulator. Of the insulators listed, dry air is the best. A high vacuum is a better insulator than dry air; however, a vacuum can be used only in special devices such as rectifier tubes. Therefore, we will be concerned in this chapter only with the more common insulators used in electrical work. We will find that conductors are usually combined with insulators; for example, a conducting wire is either covered with an insulating substance, or the conductor is fastened to insulating supports.

Table 1. Conductors and Insulators

Good Conductors	Fair Conductors	Partial Conductors	Insulators
Silver	Charcoal and coke	Water	Slate
Copper	Carbon	The body	Oils
Aluminum	Plumbago	Flame	Porcelain
Zinc	Acid solutions	Linen	Dry paper
Brass	Sea water	Cotton	Silk
Platinum	Saline solutions	Mahogany	Sealing wax
Iron	Metallic ores	Pine	Gutta percha
Nickel	Living vegetable	Rosewood	Ebonite
Tin	substances	Lignum Vitae	Mica
Lead	Moist earth	Teak	Glass
		Marble	Dry air

CONDUCTING WIRES

Wires used as electrical conductors are generally made of copper; however, aluminum is used to some extent. Silver is seldom used because of its high cost. Most wire is round, although square and rectangular forms are also used in some applications. The basic description of a round wire is its diameter. This means the diameter of the wire itself disregarding any insulation that might be used to cover the wire. The diameter of a wire can be measured accurately with a micrometer, such as illustrated in Fig. 1.

Electricians use the word *mil* to describe the diameter of a wire. A mil is 1/1000 inch. Thus, if a micrometer shows that a wire has a diameter of 0.001 inch, we say that its diameter is 1 mil. Electricians also

Fig. 1. A micrometer.

use the term *circular mil* to describe the cross-sectional area of round wire. If a wire has a diameter of 1 mil, it is said to have a cross-sectional area of 1 circular mil. A circular mil is abbreviated CM. This is a convenient way to describe area, because circular mils are equal to mils squared. For example, if a round wire has a diameter of 2 mils, its cross-sectional area is equal to 2^2, or 4 circular mils.

Because we occasionally use square or rectangular wires, we must also have a suitable way of describing the cross-sectional area of these conductors. Electricians use the term *square mil* to describe the cross-sectional area of a square or rectangular wire. If a square wire is 1 mil on a side, it is said to have a cross-sectional area of 1 square mil. It follows that if a square wire is 2 mils on a side, it has a cross-sectional area of 4 square mils. Again, if a rectangular conductor is 3 mils wide and 2 mils thick, it has a cross-sectional area of 6 square mils. Fig. 2 shows a comparison of the circular mil and the square mil.

The reason may be asked why we make a comparison of the circular mil and the square mil. If a square conductor is to be replaced by a round conductor, or vice versa, we must use the same cross-section-

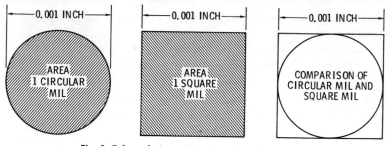

Fig. 2. Enlarged view of circular mil and square mil.

al area to obtain the same conductivity. Therefore, electricians need to know how to change circular mils into square mils, and vice versa. Note that the number of circular mils multiplied by 0.7854 gives the number of square mils. Or, the number of square mils multiplied by 1.273 gives the number of circular mils.

Thus, we write the formulas:

$$\text{Square mils} = \text{Circular mils} \times 0.7854$$

$$\text{Circular mils} = \text{Square mils} \times 1.273$$

CIRCULAR MIL-FOOT

In practical electrical work, we describe the resistance of a wire in terms of ohms per circular mil-foot. A *mil-foot* means a circular wire 1 foot in length and 1 mil in diameter. The resistance of a mil-foot of copper wire is about 10.4 ohms. Fig. 3 illustrates a circular mil-foot of copper wire. Note that the resistance of this wire is given for 20° centigrade. We will find that the resistance of a wire increases as its temperature increases. Therefore, the resistance of a circular mil-foot of wire is always listed in electrical handbooks at 20°C, or 0°C. Of course, the wire resistance is somewhat less at 0° than at 20°. We will

RESISTANCE = 10.37Ω (20°C)

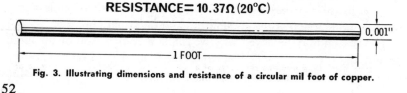

Fig. 3. Illustrating dimensions and resistance of a circular mil foot of copper.

learn how to find the resistance of a wire at any temperature that we might choose.

When we know the resistance of a wire per circular mil-foot, we can easily find the resistance for any cross section and length. The resistance per circular mil-foot is called rho (ρ) in electrical handbooks. (See Table 4.) In turn, we find the resistance of a wire of a given cross section and length by means of the formula:

$$R = \frac{\rho \times L}{CM}$$

or,

$$R = \frac{\rho \times L}{d^2}$$

where,

L is in feet,
CM is in circular mils,
d is in mils,
R is in ohms.

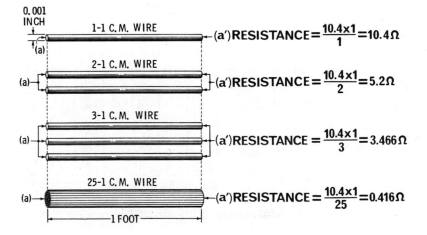

Fig. 4. Illustrating how the resistance of a conductor decreases with an increase of the area through which the current flows.

53

The foregoing formula gives the wire resistance at the temperature noted in the handbook for *rho*. We will learn how to make temperature corrections in a later part of this chapter. Fig. 4 shows how the resistance of a wire changes with a change in its cross-sectional area.

Table 2. Standard Annealed Solid Copper Wire

(American Wire Gauge—B & S)

Gauge number	Diameter (mils)	Cross section Circular mils	Cross section Square inches	Ohms per 1000 ft. 25°C. (=77° F.)	Ohms per 1000 ft. 65° C. (=149° F.)	Ohms per mile 25° C. (=77° F.)	Pounds per 1000 ft.
0000	460.0	212,000.0	0.166	0.0500	0.0577	0.264	641.0
000	410.0	168,000.0	.132	.0630	.0727	.333	508.0
00	365.0	133,000.0	.105	.0795	.0917	.420	403.0
0	325.0	106,000.0	.0829	.100	.116	.528	319.0
1	289.0	83,700.0	.0657	.126	.146	.665	253.0
2	258.0	66,400.0	.0521	.159	.184	.839	201.0
3	229.0	52,600.0	.0413	.201	.232	1.061	159.0
4	204.0	41,700.0	.0328	.253	.292	1.335	126.0
5	182.0	33,100.0	.0260	.319	.369	1.685	100.0
6	162.0	26,300.0	.0206	.403	.465	2.13	79.5
7	144.0	20,800.0	.0164	.508	.586	2.68	63.0
8	128.0	16,500.0	.0130	.641	.739	3.38	50.0
9	114.0	13,100.0	.0103	.808	.932	4.27	39.6
10	102.0	10,400.0	.00815	1.02	1.18	5.38	31.4
11	91.0	8,230.0	.00647	1.28	1.48	6.75	24.9
12	81.0	6,530.0	.00513	1.62	1.87	8.55	19.8
13	72.0	5,180.0	.00407	2.04	2.36	10.77	15.7
14	64.0	4,110.0	.00323	2.58	2.97	13.62	12.4
15	57.0	3,260.0	.00256	3.25	3.75	17.16	9.86
16	51.0	2,580.0	.00203	4.09	4.73	21.6	7.82
17	45.0	2,050.0	.00161	5.16	5.96	27.2	6.20
18	40.0	1,620.0	.00128	6.51	7.51	34.4	4.92
19	36.0	1,290.0	.00101	8.21	9.48	43.3	3.90
20	32.0	1.020.0	.000802	10.4	11.9	54.9	3.09
21	28.5	810.0	.000636	13.1	15.1	69.1	2.45
22	25.3	642.0	.000505	16.5	19.0	87.1	1.94
23	22.6	509.0	.000400	20.8	24.0	109.8	1.54
24	20.1	404.0	.000317	26.2	30.2	138.3	1.22
25	17.9	320.0	.000252	33.0	38.1	174.1	0.970
26	15.9	254.0	.000200	41.6	48.0	220.0	0.769
27	14.2	202.0	.000158	52.5	60.6	277.0	0.610
28	12.6	160.0	.000126	66.2	76.4	350.0	0.484
29	11.3	127.0	.0000995	83.4	96.3	440.0	0.384
30	10.0	101.0	.0000789	105.0	121.0	554.0	0.304
31	8.9	79.7	.0000626	133.0	153.0	702.0	0.241
32	8.0	63.2	.0000496	167.0	193.0	882.0	0.191
33	7.1	50.1	.0000394	211.0	243.0	1,114.0	0.152
34	6.3	39.8	.0000312	266.0	307.0	1,404.0	0.120
35	5.6	31.5	.0000248	335.0	387.0	1,769.0	0.0954
36	5.0	25.0	.0000196	423.0	488.0	2,230.0	0.0757
37	4.5	19.8	.0000156	533.0	616.0	2,810.0	0.0600
38	4.0	15.7	.0000123	673.0	776.0	3,550.0	0.0476
39	3.5	12.5	.0000098	848.0	979.0	4,480.0	0.0377
40	3.1	9.9	.0000078	1,070.0	1,230.0	5,650.0	0.299

AMERICAN WIRE GAUGE

Electricians usually check wire sizes with an American Standard Wire Gauge, as illustrated in Fig. 5. The American Wire Gauge is also called a Brown & Sharpe Gauge. For brevity, the term AWG or B&S is used. Gauge numbers from 0000 to 40 are listed in Table 2, with corresponding diameters in mils, cross-sectional areas in circular

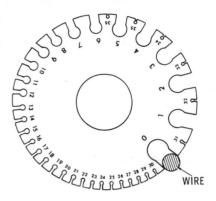

Fig. 5. Standard wire gauge.

Table 3.—Current-Carrying Capacities (in Amperes) of Single Copper
Conductors at Ambient Temperature of Below 30° C.

Size	Rubber or thermoplastic	Thermoplastic asbestos, var-cam, or asbestos var-cam	Impregnated asbestos	Asbestos	Slow-burning or weatherproof
0000	300	385	475	510	370
000	260	330	410	430	320
00	225	285	355	370	275
0	195	245	305	325	235
1	165	210	265	280	205
2	140	180	225	240	175
3	120	155	195	210	150
4	105	135	170	180	130
6	80	100	125	135	100
8	55	70	90	100	70
10	40	55	70	75	55
12	25	40	50	55	40
14	20	30	40	45	30

mils and square inches, resistance in ohms per 1000 feet at 25°C and at 65°C, resistance in ohms per mile at 25°C, and pounds per 1000 feet. These wire characteristics have been determined by the National Bureau of Standards.

In selecting a wire size for a given installation, electricians are bound by the National Electrical Code*, which establishes the allowable current capacity for insulated wires. Table 3 lists the maximum current in amperes which is permitted to flow in various sizes of wire. Rubber-insulated wire tends to heat up more than varnished cambric-insulated wire, and therefore less current is allowed in rubber-insulated wire. Other insulating materials, such as asbestos, permit more rapid escape of heat, and are thus allowed to carry more current.

STRANDED WIRES

A stranded wire consists of a group of wires which are usually twisted to form a metallic string. Stranding improves the flexibility of a wire. Fig. 6 shows a typical 37-strand conductor and how the

---

*Note:

The National Electrical Code was originally drawn in 1897 as the result of the united efforts of various insurance, electrical, architectural, and allied interests. This original Code was prepared by the National Conference on Standard Electrical Rules, composed of delgates from various interested national associations. However, since 1911, the National Fire Protection Association has been the sponsor of the National Electrical Code. The NEC establishes the minimum standards for wiring plans and installation practices in the United States. It has been adopted by the National Board of Fire Underwriters as the NBFU Regulations, and has been approved by the American Standards Association as an ASA Standard. The NEC is also the basis of more than 600 municipal electrical ordinances; however, local ordinances do not necessarily follow the NEC exactly. The NBFU has also established the Underwriter's Laboratories, Inc., which inspects electrical fittings, materials, and appliances to establish compliance with standard test specifications. Electricians may obtain the National Electrical Code and the List of Inspected Electrical Appliances from any fire underwriter's office, or from the National Board of Fire Underwriters, 85 John St., New York City, or from the Underwriters' Laboratories, Inc., 207 East Ohio St., Chicago, Illinois.

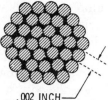

.002 INCH
37 STRAND CONDUCTOR

DIAMETER OF EACH STRAND = .002 INCH
DIAMETER OF EACH STRAND = 2 MILS
CIRCULAR MIL AREA OF EACH STRAND = D^2 = 4 CM
TOTAL CM AREA OF CONDUCTOR = 4X37 = 148 CM

Fig. 6. Stranded conductor.

total circular-mil cross section is found. We multiply the circular-mil area of each strand by the number of strands to find the total circular-mil cross section. An insulated stranded wire is called a cord. The current capacity of a cord is determined by its insulation and total circular-mil area, as for an unstranded wire.

ALUMINUM WIRE

Aluminum wire is being used to an increasing extent. Although aluminum is lighter and is easier to bend than copper, a No. 14 aluminum wire has greater resistance than an equal length of No. 14 copper wire. Therefore, larger diameter aluminum wire must be used for a given load or current demand. As an illustration, we would usually install No. 12 aluminum wire instead of No. 14 copper wire; or, we would generally install No. 4 aluminum wire instead of No. 6 copper wire. In other words, as a rough rule of thumb, we select aluminum wire two sizes larger than corresponding copper wire. However, electricians are bound by the National Electrical Code, to which reference should be made in particular situations. Again, local electrical codes are sometimes more stringent than the NEC.

LINE DROP

Since a wire has resistance, there is a voltage drop from the source end to the load end of a line. For example, if 117 volts are applied to the source end, a line drop of 20 volts would reduce the voltage at the load to 97 volts. This might be an excessive voltage drop; an electric-light bulb that normally operates at 117 volts will be dim if operated at 97 volts. To increase the load voltage, a smaller load can be used to reduce the IR drop in the line. However, if a smaller load cannot be used, we must increase the wire size. In practical situations, a reasonable compromise is determined by the electrician so that the line drop is tolerable without incurring undue cost due to the use of unnecessarily large wire.

The electrician first selects a wire size that has permissible current capacity at maximum load (maximum current flow). Then, he checks to determine whether the voltage drop will be objectionable in view of the total length of the line. If he finds that the line voltage drop will be excessive, he selects a larger size of wire to reduce the voltage drop as required. Line drop becomes the most important consideration that

Table 4. Specific Resistance

Substance	Specific resistance at 20°C.	
	Centimeter cube (microhms)	Circular-mil-foot (ohms), or ρ
Silver	1.629	9.8
Copper (drawn)	1.724	10.37
Gold	2.44	14.7
Aluminum	2.828	17.02
Carbon (amorphous)	3.8 to 4.1	
Tungsten	5.51	33.2
Brass	7.0	42.1
Steel (soft)	15.9	95.8
Nichrome	109.0	660.0

must be taken into account when a long line supplies a heavy load. Sometimes installations may be in locations where the ambient (surrounding) temperature is comparatively high, as in a furnace room. In such cases, proper allowance must be made for external heat on the allowable current flow, and each case has its own specific limitations. Maximum allowable operating temperatures are specified by the National Electrical Code.

TEMPERATURE COEFFICIENT OF RESISTANCE

We know that the resistance of a wire increases as the temperature increases. In Table 4 we see the *specific resistance* or *resisitivity* in ohms for a unit volume (the circular-mil-foot) of various metals. These specific-resistance or rho values are given at 20°C. The amount of increase in the resistance of a 1-ohm sample of a conductor per degree rise in temperature above 0°C is called its *temperature coefficient of resistance*. For copper, the temperature coefficient is approximately 0.00427. Other metals have temperature coefficients from 0.003 to 0.006.

A copper wire having a resistance of 50 ohms at 0°C will have an increase in resistance of 50 × 0.00427, or 0.214 ohms for the entire length of wire for each degree of temperature rise above 0°C. At 20°C, the increase in resistance is about 20 × 0.214, or 4.28 ohms. Thus, the total resistance at 20°C is 50 + 4.28 = 54.28 ohms. Since we must often change from Fahrenheit to centigrade temperatures, Fig. 7

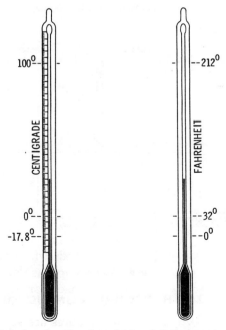

Fig. 7. Scales of centigrade and Fahrenheit thermometers.

shows the relation between these thermometer scales. To change a Fahrenheit reading to centigrade we use the formula:

$$C = \frac{5}{9}(F - 32)$$

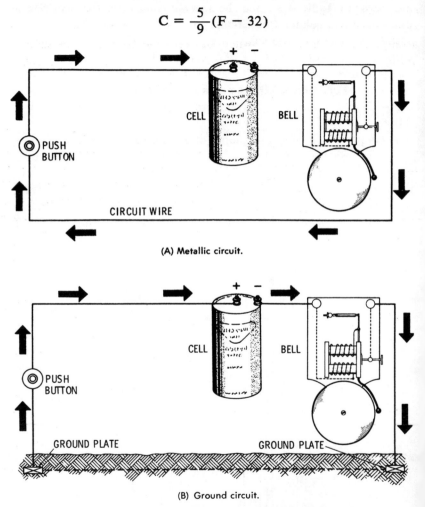

(A) Metallic circuit.

(B) Ground circuit.

Fig. 8. Simple bell circuit.

EARTH (GROUND) CONDUCTION

Moist soil is a fairly good conductor, although dry soil is a poor conductor. Therefore, moist soil can be used in case of necessity for part of a circuit. For example, electric fences for cattle (explained in detail in a later chapter) must employ a ground return circuit. In general, electricians avoid using ground circuits when possible, because there is considerable voltage drop along a ground circuit as compared with a wire circuit. To understand the principle of a ground circuit, a simple arrangement consisting of a switch, dry cell, and electric bell is shown in Fig. 8.

CONDUCTION OF ELECTRICITY BY AIR

Air is a very good insulator under most conditions. However, we know that electricity applied to a spark plug in an engine causes a spark to jump between the points of the plug. The air between the points becomes a conductor while the spark is jumping the gap. Another practical example is seen when lightning strikes from a thundercloud to the earth. The air in the path of a lightning stroke becomes a temporary conductor. We will find that air suddenly changes from an

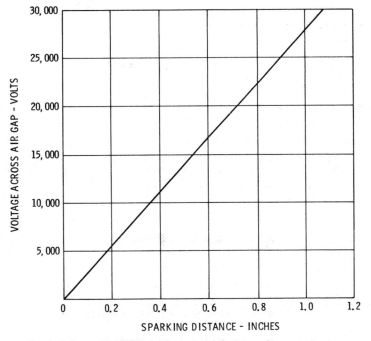

Fig. 9. Voltage required to produce a spark between sharp metal points.

insulator to a conductor at a certain critical voltage called the *sparking voltage.* Fig. 9 shows the voltage required to produce a spark between sharp metal points separated by various distances.

When a spark jumps an air gap, the resistance of the air in the path of the spark changes from an extremely high resistance to a very low resistance, such as 1 ohm or less. As soon as the spark has passed, the air ceases to be a conductor and again becomes a good insulator. It may surprise us to learn that electric currents in air are basically the same as electric currents in metals. The only difference is that the atoms in air are farther apart than the atoms in a metal. Therefore, we must apply more voltage across an air gap before electrons are transferred from one atom to the next.

Electricians work with both sparks and arcs. For example, an automotive electrician works with sparks when he adjusts and tests the spark plugs in an engine. On the other hand, he works with arcs when he welds a crack in a radiator support. The difference between a spark and an arc is that the electrodes of an arc are operated at a very high temperature, sufficient to melt iron. When air is heated to a very high temperature, its resistance becomes very low—10 ohms for example. Therefore, an arc operates at a typical voltage of 50 volts, whereas a spark operates at a typical voltage of 20,000 volts.

An arc cannot start until the electrodes are heated to a very high temperature. In practice, an electrician does this by touching one elec-

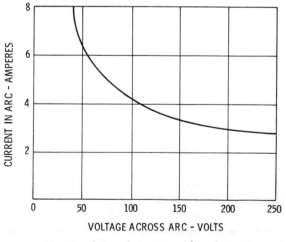

VOLTAGE ACROSS ARC - VOLTS

Fig. 10. Relation of current to voltage in an arc.

trode to the other. A heavy current flows at the point of contact, and electric power is changed into heat. The electrodes become red hot at the point of contact, and can then be separated and current will continue to flow in the extremely hot air between the electrodes. A very intense light is also produced by an arc, which will ruin the eyes unless dark goggles are worn. An arc welder operates with a spacing of about ¼ inch between iron electrodes. The voltage across the arc is about 50 volts and the current is about 6 amperes at this spacing. As the electrode spacing is changed, the voltage and current also change, as shown in Fig. 10.

Electricians who work on large searchlights are concerned with carbon arcs, as shown in Fig. 11. A carbon arc produces more flame than an iron arc, because of vaporized carbon mixed with the extremely hot air between the electrodes. The carbon electrodes are gradually con-

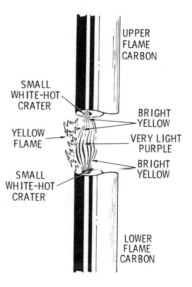

Fig. 11. The carbon arc.

sumed by vaporization, and an automatic mechanism is provided to maintain a fixed spacing between them. When a carbon electrode is nearly consumed, it must be replaced. All arcs generate intense *ultraviolet light*. Although ultraviolet light is invisible, even moderate amounts will destroy the retina of the eye. Therefore, no arc is safe to look at unless dark goggles are worn.

Gases other than air are also in wide use as conductors of electric current. For example, neon gas is used at low pressure in glass tubes to provide arc conduction and generation of light for use in advertising signs. Mercury vapor is used to provide arc conduction in bulbs or tubes that generate green-violet light. Mercury vapor is also used in automatic switches called *rectifier tubes*. A switch-type tube conducts electricity when voltage is applied in a certain polarity, but does not conduct when voltage is applied in the opposite polarity. Details of mercury-vapor arc conduction are explained in a later chapter.

CONDUCTION OF ELECTRICITY BY LIQUIDS

We know that moist soil is a fairly good conductor of electricity. It is not actually the water in the soil that provides conduction, because chemically pure water is a good insulator. Electrical conduction is provided by dissolved minerals in the water. The minerals are said to be in *water solution*. Atoms of minerals in water solution have positive or negative electric charges, and these electrically charged atoms move toward electrodes placed in the solution. Positively charged atoms move toward the negative electrode as shown in Fig. 8B, and negatively charged atoms move toward the positive electrode. This movement of charged atoms is an electric current.

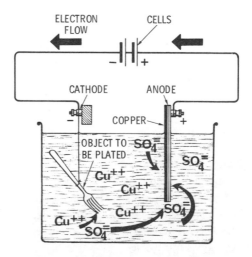

Fig. 12. How a cell may be wired for electroplating.

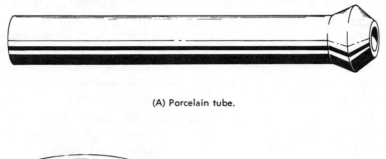

(A) Porcelain tube.

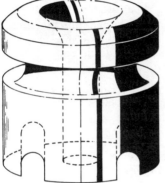

(B) Knob insulator.

(C) Spool insulator.

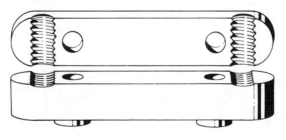

(D) Two-wire porcelain cleat.

Fig. 13. Various types of porcelain insulators.

Electricians who work with electroplating equipment are concerned with the conduction of electricity by liquids with some type of salt in solution. For example, suppose that we wish to electroplate a fork with copper, as shown in Fig. 12. A solution of blue vitriol (copper sulphate) in water is used. The fork forms the negative electrode (cathode), and a copper plate forms the positive electrode (anode). Copper sulphate dissolves to form positively charged copper atoms. Therefore, the charged copper atoms are attracted to the negative fork, and a film of metallic copper is deposited on the fork.

If we wish to silver plate a fork, we will use a solution of silver cyanide. Silver cyanide is a poisonous silver salt which dissolves in water to form positively charged silver atoms. Thus, the electroplating process is basically the same as in the case of a copper-sulphate solution. If we wish to gold plate a fork, we will use a solution of gold cyanide. Gold cyanide is also a poisonous salt. Therefore appropriate care must be observed by electricians who work with electroplating equipment containing cyanides. Even a small amount of cyanide can be deadly if taken into the human body.

INSULATORS FOR SUPPORT OF WIRES

Many types of insulators are used to support electrical wires. Porcelain insulators with forms such as those shown in Fig. 13 are in very

Fig. 14. Porcelain tubes inserted in wall studs.

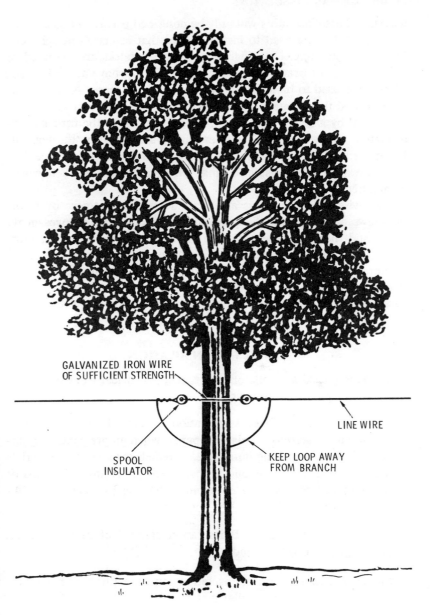

GALVANIZED IRON WIRE
OF SUFFICIENT STRENGTH

LINE WIRE

SPOOL
INSULATOR

KEEP LOOP AWAY
FROM BRANCH

Fig. 15. Insulation of a line wire from a tree.

wide use. These insulators will only be described briefly in this chapter. Porcelain tubes are used to insulate wires that are run through holes in walls or studding. (See Fig. 14). Knob insulators are mounted on walls, ceilings, or beams by nails or screws. The groove around a knob insulator is used to bind the line wire securely in place. A spool insulator is used to insulate a wire from a wall. The insulator is typically suspended by a length of stout wire running through the center of the insulator, and secured to a screw-eye in the wall. In turn, the line wire is bent around the groove on the insulator and secured in place. Fig. 15 shows how spool insulators may be used to insulate a line wire from a tree.

Two-wire porcelain cleats are secured to wood surfaces by means of screws or nails. The line wires are gripped by the grooves in the cleats and thereby prevented from sliding. Cleat-wiring installations can be made in a comparatively short time. However, the type of insulators used in various situations and locations are governed by regulations which must be observed by the electrician. Details are discussed later under the topics of house wiring and power wiring.

CLASSES OF INSULATION

Substances used as insulators in practical electrical work are classified into four groups, as follows:

Class A Insulation. Class A insulation consists of: (1) cotton, silk, paper, and materials similar to paper when impregnated or immersed in an insulating liquid; (2) molded or laminated materials with cellulose filler, phenolic resins, or similar resins; (3) films or sheets of cellulose acetate or similar cellulose products; and (4) varnishes or enamel applied to conductors.

Class B Insulation. Class B insulation consists of mica, asbestos, or fiberglass, all with a binder.

Class C Insulation. Class C insulation consists entirely of mica, porcelain, glass, quartz, or similar materials.

Class O Insulation. Class O insulation consists of cotton, silk, paper, or similar materials that are *not* impregnated or immersed in an insulating liquid.

We say that an insulation is *impregnated* when the air spaces within the insulation are filled up by an impregnating substance such as paraffin. An impregnating substance is itself a good insulator.

INSULATION RESISTANCE

Since no insulator is perfect, the insulation resistance of an insulated conductor may be measured in megohms. An instrument called a *megger* is used when it is desirable or necessary to measure insulation resistance. A megger is a type of ohmmeter that measures resistance in megohms (millions of ohms), and can measure up to several thousand megohms. Any insulating substance will break down at some value of high voltage, and the insulation of commercial insulated conductors is rated (guaranteed) to withstand a certain value of voltage. Therefore, meggers are designed to measure insulation resistance at high voltages.

PLASTIC-INSULATED SHEATHED CABLES

Plastic-insulated sheathed cables such as illustrated in Fig. 16 are used extensively for residential and commercial wiring. The indoor type plastic sheathed cable has a tough and flexible outer

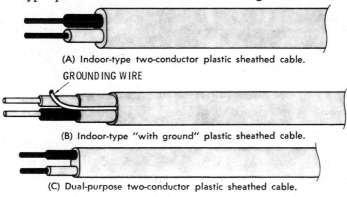

(A) Indoor-type two-conductor plastic sheathed cable.

GROUNDING WIRE

(B) Indoor-type "with ground" plastic sheathed cable.

(C) Dual-purpose two-conductor plastic sheathed cable.

Fig. 16. Typical plastic-insulated sheathed cable.

jacket which is ivory colored and flat in shape. It is installed for indoor wire runs. The inner insulation is heavy thermo-plastic. Two-conductor cable is used in "old work", when the existing cable is of the two-conductor type. "New work" consisting of wiring in new buildings must use grounded-type receptacles; therefore, "with-ground" type cable must be utilized. Note that dual-purpose plastic sheathed cable can be installed underground outdoors, or used for indoor wire runs. It can usually be buried without conduit (protective metallic tubing) unless there is a possibility of mechanical damage. Dual-purpose plastic sheathed cable resists moisture, acids, and corrosion. It can be run through masonry or between studding. If installed for outdoor circuits, "with ground" type of dual-purpose cable must be used.

SUMMARY

A conductor is a material or substance that carries electric current. No conductor is perfect, and because all conductors have some small amount of resistance, it is better to define a conductor as a substance with a very low resistance. An insulator such as slate, paper, and glass is a substance that does not permit electric current to flow easily.

Wires used as an electric conductor are generally made of copper or aluminum. Most wire is round and the diameter can be measured accurately with a micrometer. Electricians use the word *mil* to describe the diameter of wire. Circular mil is used to describe the cross-sectional area of round wire. Square or rectangular wire is used in some applications and its cross-sectional area is measured in square mils.

In practical electrical work, we describe the resistance of a wire in terms of ohms per circular wire 1 foot in length and 1 mil in diameter. When we know the resistance of a wire per circular mil-foot, we can easily find the resistance for any cross-section and length. The resistance per circular mil-foot is called *rho*.

Since a wire has resistance, there is a voltage drop from the source end of the wire to the load end of the wire. Proper wire size is selected to permit current capacity at maximum load to flow. If the line voltage drop will be excessive, a larger size wire is used to reduce the voltage drop. Line drop becomes the most important consideration that must be taken into account.

TEST QUESTIONS

1. What is the definition of a conductor? Of an insulator?
2. Give an example of a good conductor, a fair conductor, a partial conductor, and an insulator.
3. What is a micrometer?
4. Define a mil.
5. How are mils related to circular mils?
6. Compare a circular mil to a square mil.
7. Why are electricians concerned with circular mils and square mils?
8. Explain the meaning of circular-mil-foot of wire.
9. Does the resistance of copper wire increase or decrease as the temperature increases?
10. Is the AWG wire gauge the same as the B&S gauge?
11. How is the resistance of a wire related to its cross-sectional area? To its length?
12. Describe the use of an American Standard Wire Gauge.
13. What is meant by the current-carrying capacity of a given wire size?
14. Why are stranded conductors used in lamp cords?
15. Explain the meaning of line drop.
16. Is the resistance of a circular-mil-foot (specific resistance) of copper wire greater or less than that of aluminum wire?
17. Give a general comparison of the Fahrenheit and centigrade thermometer scales.
18. How do we change a Fahrenheit temperature into a centigrade temperature?
19. Describe the meaning of a ground-return circuit.
20. About how much voltage can be withstood by needle points ½ inch apart without breakdown and sparking through the air between the points?
21. In what way does an arc differ from a spark?
22. How does a salt solution conduct electricity?
23. Give several examples of insulators used to support line wires.

CHAPTER 3

Electric Circuits

We have learned that an electric circuit is a closed path for current flow. Electricians are concerned with many types of circuits. It has been noted in the first chapter that a *series circuit* is a circuit that supplies electricity to one or more loads; all devices in a series circuit are connected end-to-end in a closed path and the same amount of current flows through each device. We are now ready to consider some more facts about series circuits which are of practical importance to an electrician.

PICTURE DIAGRAMS AND SCHEMATIC DIAGRAMS

Fig. 1 shows a picture diagram of two electric lamps connected in series with a battery. We occasionally use picture diagrams, but generally work from schematic diagrams. A schematic diagram shows an electric circuit by means of *graphical symbols* instead of outline pictures. For example, Fig. 2 shows two schematic diagrams. Standard electrical symbols are used to represent a battery, a lamp, and a resistor. Next, let us observe that the schematic diagram in Fig. 2A can be represented by an *equivalent circuit,* as seen in Fig. 2B. An

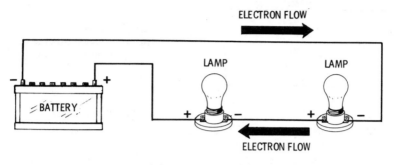

ELECTRON FLOW

LAMP   LAMP

BATTERY

ELECTRON FLOW

Fig. 1. Picture diagram of two electric lamps connected in series with a battery.

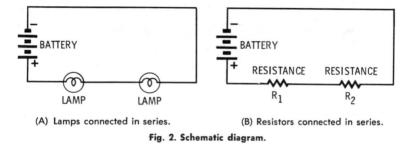

BATTERY   BATTERY

RESISTANCE   RESISTANCE

LAMP   LAMP R_1 R_2

(A) Lamps connected in series.   (B) Resistors connected in series.

Fig. 2. Schematic diagram.

equivalent circuit has the same electrical properties as the original circuit, but an equivalent circuit does not serve the same purpose. Let us see what this means.

If R_1 and R_2 in Fig. 2B have the same amounts of resistance as the lamps in Fig. 2A, and if the batteries in both circuits have the same voltage, it is clear that each circuit will draw the same amount of current. However, these circuits do not serve the same purpose, because the circuit in Fig. 2A produces light and heat, while the circuit in Fig. 2B produces heat only. Nevertheless, the equivalent circuit is useful because it is the first step in *reducing* the original series circuit into a simplified circuit.

It is evident that the load in Fig. 2B consists of resistance R_1 plus resistance R_2. Therefore, we can combine R_1 and R_2 into a single resistor, as seen in Fig. 3. For example, if each lamp in Fig. 2A has a resistance of 25 ohms, then $R_1 = 25$ ohms and $R_2 = 25$ ohms in Fig. 2B. In turn, $R_L = 50$ ohms in Fig. 3. The circuit in Fig. 3 represents the final reduction, or simplest possible equivalent circuit for the lamp

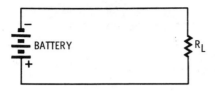

Fig. 3. Equivalent circuit of resistor
connected in series.

circuit in Fig. 2A. Equivalent circuits are useful because they help us to better understand the operation of complicated circuits.

VOLTAGE POLARITIES IN SERIES CIRCUITS

Electricians must understand voltage polarities in series circuits. Therefore, let us observe the polarities shown in Fig. 1. Electrons flow from the negative terminal of the battery to the right-hand lamp in the diagram. Of course there is a *voltage drop* across this lamp. Since the right-hand terminal of the lamp is connected to the negative terminal of the battery, and the left-hand terminal of the lamp eventually returns to the positive terminal of the battery, it is obvious that the right-hand terminal of the lamp has negative polarity with respect to its left-hand terminal.

The same amount of current flowing through the right-hand lamp also flows through the left-hand lamp in Fig. 1. Using the same reasoning as before, we see that the right-hand terminal of this lamp must be negative with respect to its left-hand terminal. The voltage drops across the two lamps are in *series-aiding,* and these two voltage drops add up to the same amount of voltage as we find across the battery terminals. This is an example of *Kirchhoff's voltage law,* which was noted in the first chapter.

To show why we need to observe the polarity of a voltage, let us consider the connection of a voltmeter across a battery, as shown in Fig. 4. To measure the battery voltage, we must connect the positive terminal of the voltmeter to the positive terminal of the battery, and we must connect the negative terminal of the voltmeter to the negative terminal of the battery. The pointer will move up-scale on the voltmeter. If we make a mistake and reverse the polarity of the meter connections, the pointer will not move up-scale on the voltmeter; instead, the pointer will move off-scale to the left. Therefore, voltmeters have their positive and negative terminals indicated, as seen in Fig. 5.

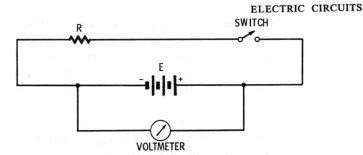

Fig. 4. A voltmeter connected across a battery in a series circuit.

Courtesy Simpson Electric Co.

Fig. 5. The voltmeter has its terminal polarity marked.

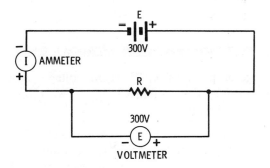

Fig. 6. An ammeter and voltmeter connected into a circuit.

An ammeter also has its positive and negative terminals indicated; an ammeter must be connected into a circuit in proper polarity, as shown in Fig. 6. The voltmeter is connected to read the voltage drop

75

across R. Note carefully that the ammeter must *not* be connected across R; in effect, the battery would thereby be short-circuited through the ammeter. The ammeter would probably be damaged. Even if the ammeter did not burn out, the short circuit would soon ruin the battery. Therefore, we must keep the following rules in mind:

1. An ammeter is always connected in *series* with a circuit.
2. A voltmeter is always connected *across* the battery, resistor, or other device to measure a voltage drop.

VOLTAGE MEASUREMENTS WITH RESPECT TO GROUND

A series circuit with a ground return is shown in Fig. 7. The circuit consists of battery E and resistors R_1, R_2, and R_3. Voltmeter V_1 mea-

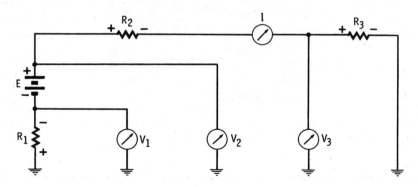

Fig. 7. Voltage measurements with respect to ground.

sures the voltage drop across R_1; V_2 is the voltage across the *battery and R_1*. Note carefully that to find voltage E, we *subtract* voltage V_1 from voltage V_2; we write:

$$E = V_2 - V_1$$

Next, we observe that the voltage drop across R_2 is found by subtracting V_3 from V_2. Finally, V_3 is the voltage drop across R_3. For example, if $R_1 = 1$ ohm, $R_2 = 2$ ohms, $R_3 = 3$ ohms, and $E = 6$ volts, the current flow $I = 1$ ampere. In turn, $V_1 = 1$ volt, $V_2 = 5$ volts, and $V_3 = 3$ volts. Note that we can disregard the voltage drop across ammeter I in Fig. 7 because the resistance of an ammeter is very small.

Note also in Fig. 7 that the letters E and V are used to indicate voltages. We commonly use E to indicate a *source voltage,* and use V to indicate a *voltage drop.* This is helpful, because it prevents confusion when we are working out a circuit problem. However, many electricians simply use E to indicate any voltage in a circuit, and other electricians use V to indicate any voltage in a circuit. It does not make any difference whether we write V or E, or both, as long as we remember which voltage the letter stands for.

RESISTANCE OF A BATTERY

Any battery has a certain amount of resistance, called its *internal resistance.* In many circuits, we can disregard the internal resistance of a battery without getting into practical difficulties. On the other hand, we will find various circuits in which it is necessary to take the internal resistance of a battery into account. Fig. 8A illustrates the internal construction of a dry cell. The paste contains a chemical solution called an *electrolyte.* This electrolyte has a small amount of resistance in a brand-new cell, and in a nearly dead cell the resistance will be very great. Therefore, any cell can be represented by the equivalent circuit shown in Fig. 8B. The cell operates electrically as if a perfect cell E were connected in series with an internal resistance R_{in}.

When a cell is connected to a load resistor, as shown in Fig. 8C, the voltage across R_L will be less than E. There is a voltage drop inside the cell equal to IR_{in}. Therefore, the voltage across R_L is equal to $E - IR_{in}$. We call E the *electromotive force,* or *emf* of the cell. On the other hand, the voltage drop across R_L in Fig. 8C is equal to the *terminal voltage* of the cell. We will find that E remains about the same (1.5 volts) even in a nearly dead cell. However, as noted previously, the resistance of R_{in} becomes greater as the cell becomes weaker. It follows that the terminal voltage of a cell is the same as its emf when no current is being drawn, even if the cell is nearly dead.

Therefore, we cannot test a cell properly with a voltmeter alone, because a voltmeter draws a very small amount of current. Instead, we

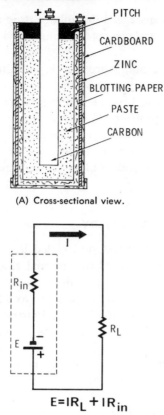

(A) Cross-sectional view.

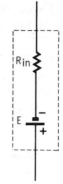

(B) Equivalent circuit.

(C) Load resistor connected to equivalent circuit.

$$E = IR_L + IR_{in}$$

Fig. 8. Internal resistance of a dry cell.

must connect a load resistance across the cell as shown in Fig. 8C and measure the voltage drop across the load resistor. A large dry cell has more electrode area than a small flashlight cell. A large cell can normally supply more current than a small cell. This means that a large cell should be tested with a load resistor that has a comparatively small value. Standard battery testers, such as illustrated in Fig. 9, consist of a voltmeter and a dozen load resistors with different resistance values.

The voltage-selector switch on a battery tester connects a load resistor across a battery that provides a normal load or current drain. For example, an ordinary No. 6 dry cell should maintain a terminal

78

Fig. 9. A commercial battery tester.

Courtesy Simpson Electric Co.

voltage of 75% of 1.5 volts, or at least 1⅛ volts at a current drain of 10 amperes. This is called the *intermittent duty* rating of a dry cell. A dry cell is intended to supply a current of 10 amperes only for a short time and at widely separated intervals. On the other hand, the *constant-current* rating of a dry cell is the amount of current that the cell is intended to supply steadily over a period of several hours. In the case of a No. 6 dry cell, the constant-current rating is ⅛ ampere.

Ordinary flashlights use a size-D dry cell; since it is much smaller than a No. 6 cell, a size-D cell has an intermittent duty rating of 2 amperes. Hearing aids use very small cells which can supply much less current, either on intermittent duty or in steady service. The internal resistance of a No. 6 dry cell in good condition is about 0.1 ohm. As the cell becomes weaker, its internal resistance gradually increases. When a cell is so weak that it reads "bad" on a battery tester, its internal resistance has increased greatly. However, the emf of a bad cell will be practically the same as that of a brand-new cell.

79

EFFICIENCY AND LOAD POWER

Efficiency means output power divided by input power. For example, if we get half as much power out of a machine as we put into it, we say that the efficiency of the machine is 50%. The same principle applies to electric circuits. The *power* that we get out of the circuit shown in Fig. 10 is the *power* in the lamp filament R_L; the power that

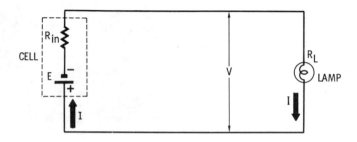

Fig. 10. The circuit efficiency depends on the value of R_{in}.

goes into the circuit is supplied by the cell. We know that the power in the lamp is equal to VI watts, and the power supplied by the cell is equal to EI watts. Therefore, the efficiency of the circuit is equal to VI/EI, or V/E. In terms of percentage, the efficiency of the circuit is equal to 100V/E percent.

The efficiency of the circuit in Fig. 10 could be 100% only if R_{in} were zero. Since R_{in} can never be zero, the circuit efficiency must always be less than 100%. We will find that the efficiency of a circuit is greatest when the load is very light—that is, when a very small amount of current is drawn by the load. Let us see why this is so. If R_L has a very high value, then IR_L is much greater than IR_{in}. This is just another way of saying that V is large when R_L is large. Since E remains the same, it follows that the circuit efficiency V/E is high when R_L has a high value.

Let us consider how we can get the greatest power out of the circuit shown in Fig. 10. We will find the value of R_L that makes VI as large as possible. It can be shown that the lamp will have the largest number of watts when $R_L = R_{in}$. This is a surprising answer at first glance. Let us note the following facts:

1. If R_L were zero, maximum current would flow in the circuit. However, V would then be zero, and the power in the load would be zero.

2. If R_L were infinite, maximum voltage would be dropped across the load. However, I would then be zero, and the power in the load would be zero.

3. The load power has its greatest value when the load resistance is equal to the internal resistance of the cell.

A practical example is shown in Fig. 11. The source voltage is 100 volts, and the source resistance is 5 ohms. As the resistance of the load is changed, the load voltage, circuit current, load power, and circuit efficiency change as shown in Fig. 11B. When the load resistance is equal to the source resistance (5 ohms), the load power has its greatest value. If the load resistance is less or greater than 5 ohms, the load power decreases. We say that *maximum power transfer* occurs when the load resistance is equal to the source resistance. Note that the circuit efficiency is only 50% when maximum power transfer is obtained. Fig. 11C shows how the efficiency, load voltage, circuit current, and load power change as the load resistance is changed.

CIRCUIT VOLTAGES IN OPPOSITION

Up to now we have only considered cells connected in series-aiding. However, in practical electrical work, cell voltages may be connected in series-opposing. Fig. 12 shows a 1.5-volt source connected in series-opposing with a 3-volt source. The 1.5 volts subtracts from the 3 volts, and the voltmeter reads 1.5 volts. Note the voltmeter polarity. Automobile electricians are concerned with series-opposing voltages in storage-battery charging circuits. Therefore, let us briefly consider the properties of storage batteries and battery-charging circuits.

A storage battery is also called a secondary battery, whereas a dry cell is called a primary battery. Secondary cells are different from primary cells in that a secondary cell can be recharged, whereas a primary cell cannot. The basic storage cell consists of a pair of lead plates immersed in a solution of sulfuric acid. Fig. 13 depicts a simple storage cell and charging circuit. An electric generator is used to supply the

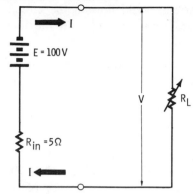

E = OPEN-CIRCUIT VOLTAGE OF SOURCE
R_{in} = INTERNAL RESISTANCE OF SOURCE
V = TERMINAL VOLTAGE
R_L = RESISTANCE OF LOAD
P_L = POWER USED IN LOAD
I = CURRENT FROM SOURCE
% EFF = PERCENTAGE OF EFFICIENCY

(A) Circuit.

R_L	V	I	P_L	% EFF
0	0	20	0	0
1	16.6	16.6	267.6	16.6
2	28.6	14.3	409	28.6
3	37.5	12.5	468.8	37.5
4	44.4	11.1	492.8	44.4
5	50	10	500	50
6	54.5	9.1	495.4	54.5
7	58.1	8.3	482.2	58.1
8	61.6	7.7	474.3	61.6
9	63.9	7.1	453.7	63.9
10	66	6.6	435.6	66
20	80	4	320	80
30	87	2.9	252	87
40	88	2.2	193.6	88
50	91	1.82	165	91

(B) Chart.

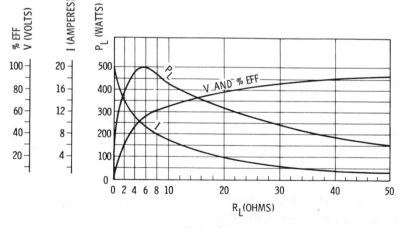

(C) Graph.

Fig. 11. Effect of source resistance on power output.

charging voltage and current. Before the storage cell is charged, it has very little resistance, and the circuit current is practically 3 amperes.

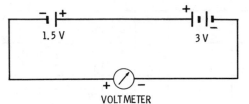

Fig. 12. Circuit voltages in opposition.

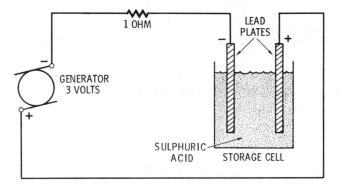

Fig. 13. A simple storage cell and charging circuit.

As the charging process continues in Fig. 13, chemical changes take place at the surfaces of the lead plates, and the storage cell builds up a voltage between its terminals. This voltage has the polarity shown in the diagram, and opposes the generator voltage. Therefore, the charging current becomes less as the storage cell charges. Fig. 14 shows how the terminal voltage of a storage cell increases during charge. The terminal voltage rises to 2.2 volts, then remains comparatively constant for a time, and finally rises to a maximum of 2.6 volts. At this point, the effective voltage in the circuit of Fig. 13 is $3 - 2.6$ $= 0.4$ volt.

When a storage cell is removed from the charging circuit, its terminal voltage falls to approximately 2 volts; the cell can then be used as a voltage source until the chemical film on its plates has been used up. The voltage of the cell remains at practically 2 volts until it is almost discharged, and then falls rapidly. After the cell is discharged, it can be recharged as has been explained. We measure the *capacity* of a storage cell in ampere-hours. Its ampere-hour capacity is equal

83

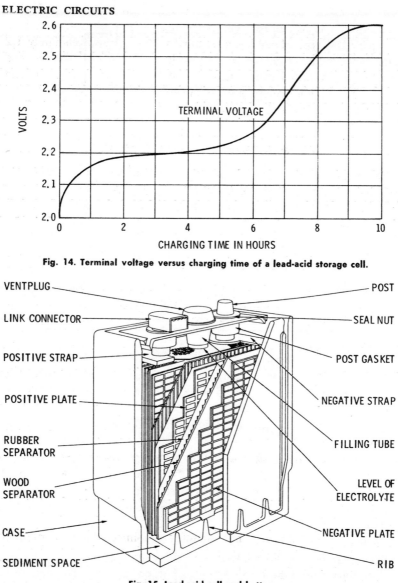

Fig. 14. Terminal voltage versus charging time of a lead-acid storage cell.

Fig. 15. Lead-acid cell and battery.

to the number of amperes supplied by the cell on discharge, multiplied by the number of hours that the cell can supply current. Commercial storage cells and batteries have the construction shown in Fig. 15.

Another type of storage cell, called the *nickel-cadmium* cell, is in wide use. This type of cell has a terminal voltage of approximately 1.2 volts; however, it is not as heavy and has a longer life than a lead storage cell. A nickel-cadmium cell also requires less attention and care; it can be completely discharged and left uncharged for an indefinite time. This abusive treatment would ruin a lead cell.

PRINCIPLES OF PARALLEL CIRCUITS

Electricians work extensively with parallel circuits. In a parallel circuit, each load is connected in a *branch* across the voltage source as shown in Fig. 16. Therefore, there are as many paths for current flow

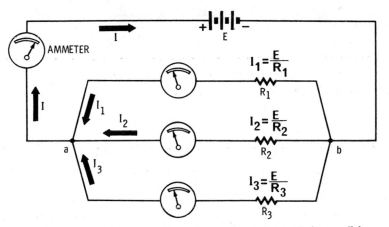

Fig. 16. Illustrating a circuit having three resistances connected in parallel.

as there are branches. The loads in a parallel circuit are sometimes said to be connected in *multiple*. We know that when more loads are connected into a series circuit, the total circuit resistance increases. On the other hand, when more loads are connected into a parallel circuit, the total resistance decreases.

We observe in Fig. 16 that the voltage across all branches of a parallel circuit (from *a* to *b*) is the same because all branches are connected directly to the voltage source E. Each load resistor draws current independently of the other load resistors. Each branch current depends only on the load resistance in that branch. Therefore, we find

85

the current in each branch by dividing its load resistance into the source voltage. The application of Ohm's law in each branch is shown in Fig. 16. Since the source voltage E is applied across each branch, it follows that we may write:

$$E = I_1 R_1 = I_2 R_2 = I_3 R_3$$

Also, the three branch currents must be supplied by the total current I in Fig. 16. Therefore, we write:

$$I = I_1 + I_2 + I_3$$

$$\text{or,} \quad I = \frac{E}{R_1} + \frac{E}{R_2} + \frac{E}{R_3}$$

Next, let us find an equivalent circuit for the parallel circuit in Fig. 16. We will replace R_1, R_2, and R_3 with an equivalent resistor R_{eq}. Since $I = E/R_{eq}$, we will write:

$$\frac{E}{R_{eq}} = \frac{E}{R_1} + \frac{E}{R_2} + \frac{E}{R_3}$$

We observe that E cancels out in the foregoing formula, leaving:

$$\frac{1}{R_{eq}} = \frac{1}{R_1} + \frac{1}{R_2} + \frac{1}{R_3}$$

To show how the foregoing formula is used, let us take a practical example. Suppose that $R_1 = 5$ ohms, $R_2 = 10$ ohms, and $R_3 = 30$ ohms in Fig. 16. Then, we find the value of R_{eq} as follows:

$$\frac{1}{R_{eq}} = \frac{1}{5} + \frac{1}{10} + \frac{1}{30} = \frac{10}{30} = \frac{1}{3}$$

or,

$$\frac{1}{R_{eq}} = 0.2 + 0.1 + 0.033 = 0.333$$

Therefore,

$$R_{eq} = 3 \text{ ohms, approximately}$$

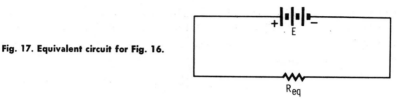

Fig. 17. Equivalent circuit for Fig. 16.

Our equivalent circuit is drawn as shown in Fig. 17, and the value of R_{eq} is approximately 3 ohms. This is the method used by most electricians to find the equivalent resistance of several resistors connected in parallel. It makes no difference how many branches there may be in a parallel circuit; we simply add up the reciprocals of all the load resistors to find the reciprocal of the equivalent resistance. Then, the equivalent resistance is the denominator of this fraction.

SHORT CUTS FOR PARALLEL CIRCUITS

Electricians use certain short cuts in solving parallel circuits. The first short cut applies to any number of parallel resistors, provided each resistor has the same resistance value. In this case, the equivalent resistance is found by dividing the resistance of one resistor by the number of resistors that are connected in parallel. For example, suppose that we have five 10-ohm resistors connected in parallel as shown in Fig. 18. In turn, we write:

$$R_{eq} = \frac{10}{5} = 2 \text{ ohms}$$

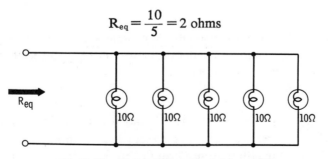

Fig. 18. Five 10-ohm loads connected in parallel.

The second short cut applies when we have two resistors with different values connected in parallel. In this case, the equivalent resistance is equal to the product of the two resistances divided by their

87

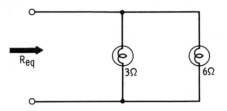

Fig. 19. Two loads of unequal value connected in parallel.

sum. For example, if we have two resistors with values of 3 ohms and 6 ohms connected in parallel as shown in Fig. 19, their equivalent resistance is found as follows:

$$R_{eq} = \frac{R_1 R_2}{R_1 + R_2} = \frac{3 \times 6}{3 + 6} = \frac{18}{9} = 2 \text{ ohms}$$

An easy way to find the resistance of a parallel circuit is shown in Fig. 20. This is called the leaning-ladder method; it is a graphical solution. In this example, $R_1 = 6$ ohms, and $R_2 = 12$ ohms. We draw the corresponding lines as shown, and note that their intersection occurs at the 4-ohm level. Therefore, the resistance of the parallel combination of R_1 and R_2 is 4 ohms. Note that the values indicated in Fig. 18 may all be multiplied by 10, as shown in Fig. 20. In turn, we can solve parallel circuits that have higher resistance values. Thus, if $R_1 = 100$ ohms, and $R_2 = 25$ ohms, the resistance of the parallel combination is 20 ohms. Note that the values indicated in Fig. 21 may all be multiplied by 10, to solve parallel circuits that have still higher resistance values.

It is also easy to find the resistance of three parallel-connected resistors, as shown in Fig. 22. In this example, $R_1 = 10$ ohms, $R_2 = 8$ ohms, and $R_3 = 6$ ohms. We make the graphical solution in two steps. First, we draw the lines for R_1 and R_2, which intersect at point P. Then, we draw a line from O on OP to 6 on the lefthand vertical axis. This gives us the point of intersection P_1, showing that the parallel resistance of R_1, R_2, and R_3 is 2.55 ohms, approximately. The same method may be extended to any number of resistors connected in parallel.

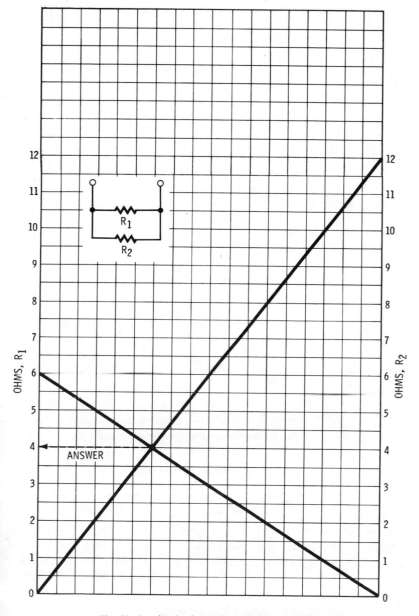

Fig. 20. Graphical solution for resistors in parallel.

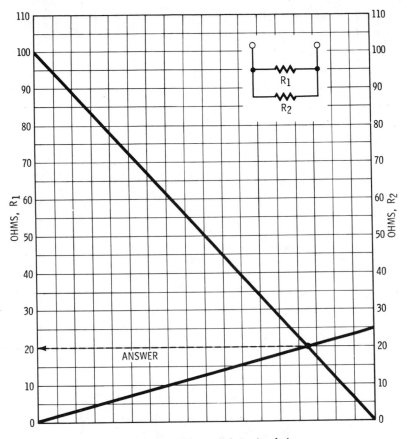

Fig. 21. Another parallel circuit solution.

CONDUCTANCE VALUES

Some electricians prefer to work with conductance values instead of resistance values when solving parallel circuits. Conductance is equal to $1/R$. We measure conductance in mhos, whereas we measure resistance in ohms. In other words, conductance is the reciprocal of resistance, and mho is the reciprocal of ohm. We let G stand for conductance, whereas we let R stand for resistance. The equivalent or total conductance of a parallel circuit is equal to the reciprocal of its equivalent or total resistance. Thus, we write:

90

$$G_{eq} = \frac{1}{R_{eq}}$$

or,

$$R_{eq} = \frac{1}{G_{eq}}$$

It is easy to see that the equivalent conductance of a parallel circuit is equal to the sum of its branch conductances. For example, the equivalent conductance of the circuit shown in Fig. 23 is written:

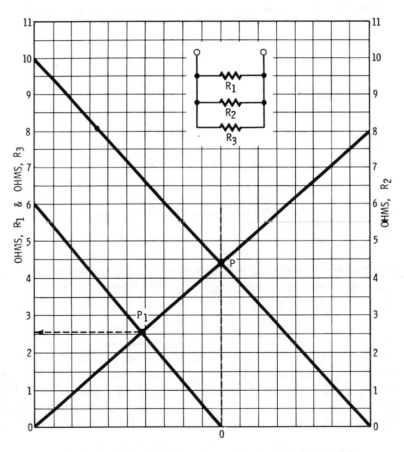

Fig. 22. Graphical solution for three resistors connected in parallel.

$$G_{eq} = G_1 + G_2 + G_3 + G_4 + G_5$$

Since G and R are reciprocals of each other, their Ohm's-law equivalents are also reciprocals of each other. In turn, we write:

$$R = \frac{E}{I}$$

$$G = \frac{I}{E}$$

Therefore, Ohm's law is written with conductance values as follows:

$$I = EG$$

$$E = \frac{I}{G}$$

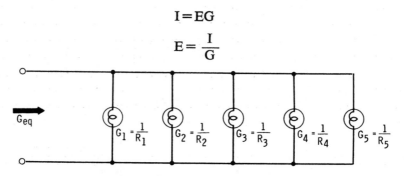

Fig. 23. Five conductances connected in parallel.

It seems more difficult at first to use conductance values instead of resistance values in solving parallel circuits. However, after we become familiar with the use of conductance values, we find it easier to solve complicated parallel circuits.

If an electrician prefers to use a graphical solution to convert from resistance to conductance, he may use the method shown in Fig. 24. Note that the horizontal axis is laid off in steps from 0 to 1; the vertical axis is laid off in steps from 0 to 9. A line has been drawn at the unit level at the vertical axis for reference. To find the conductance of 5 ohms, we draw a line from 0 to 5, as shown. This line intersects the unit level at point P, which shows that the conductance value is 0.2 mho. Of course, we could also convert from conductance to resistance values by reversing the graphical procedure.

92

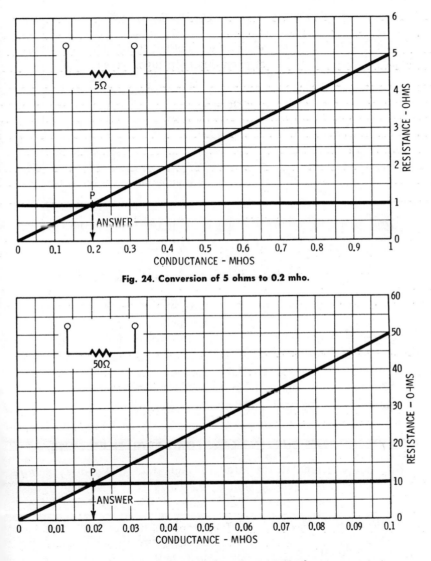

Fig. 24. Conversion of 5 ohms to 0.2 mho.

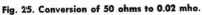

Fig. 25. Conversion of 50 ohms to 0.02 mho.

Note that the values along the horizontal axis may be divided by 10, and the values along the vertical axis multiplied by 10. Thus, a resistance of 50 ohms corresponds to a conductance of 0.02 mho. Or,

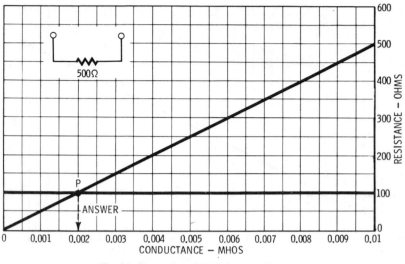

Fig. 26. Conversion of 500 ohms to 0.002 mho.

the values along the horizontal axis may be divided by 100, and the values along the vertical axis multiplied by 100. For example, a resistance of 500 ohms has a conductance of 0.002 mho. This procedure is illustrated in Fig. 25, wherein horizontal values have been divided by 10, and vertical values multiplied by 10. Again, in Fig. 26, horizontal values have been divided by 100, and vertical values multiplied by 100.

KIRCHHOFF'S CURRENT LAW FOR PARALLEL CIRCUITS

Kirchhoff's current law for parallel circuits states that *at any junction of conductors, the algebraic sum of the currents is zero.* This is just another way of saying that just as much electricity must leave a junction as there is electricity entering the junction. For example, let us consider junction *a* in Fig. 27. We assume that the current flowing toward junction *a* is positive, and that the currents flowing away from junction *a* are negative. We consider that I_t is positive, and that I_1, I_2, and I_3 are negative. In turn, we write Kirchhoff's current law:

$$+I_t - I_1 - I_2 - I_3 = 0$$

94

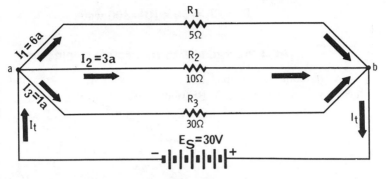

Fig. 27. Resistors in parallel.

Or, for the example shown in Fig. 27, we write:

$$+ 10 - 6 - 3 - 1 = 0$$

The importance of Kirchhoff's current law will be shown later by means of practical examples. We will find that, just as in a series circuit, the total power consumed in a parallel circuit is equal to the sum of the power values in each resistor. For example, the power P_1 in R_1 of the circuit in Fig. 27 is written:

$$P_1 = EI_1 = 30 \times 6 = 180 \text{ watts}$$

Next, power P_2 in R_2 is found:

$$P_2 = EI_2 = 30 \times 3 = 90 \text{ watts}$$

Similarly, power P_3 in R_3 is found:

$$P_3 = EI_3 = 30 \times 1 = 30 \text{ watts}$$

The total power consumed by the parallel circuit in Fig. 27 is simply the sum of the power values in each branch:

$$P_t = P_1 + P_2 + P_3 = 180 + 90 + 30 = 300 \text{ watts}$$

We can check this answer by using the total current value in the power formula:

95

$$P_t = EI_t = 30 \times 10 = 300 \text{ watts}$$

PRACTICAL PROBLEMS IN PARALLEL CIRCUITS

Let us consider the parallel circuit shown in Fig. 28. This circuit has two branches, *a* and *b*. Branch *a* has 3 lamps in parallel. L_1 takes a power of 50 watts, L_2 takes 25 watts, and L_3 takes 75 watts. Branch

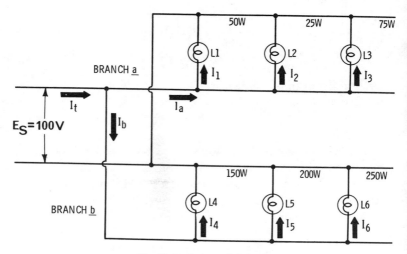

Fig. 28. Typical parallel circuit.

b also has 3 lamps in parallel; L_4 takes 150 watts, L_5 takes 200 watts, and L_6 takes 250 watts. The source voltage is 100 volts. Our problem is to:

1. Find the current in each lamp.
2. Find the resistance of each lamp.
3. Find the current in branch *a*.
4. Find the current in branch *b*.
5. Find the total circuit current.
6. Find the total circuit resistance.
7. Find the total power in the circuit.

This problem is solved by means of Ohm's law, the power law, and Kirchhoff's current law. We proceed as follows:

1. The current in L_1 is $I_1 = P_1/E_s = 50/100 = 0.5$ ampere. Similarly, $I_2 = 0.25$ ampere, $I_3 = 0.75$ ampere, $I_4 = 1.5$ amperes, $I_5 = 2$ amperes, and $I_6 = 2.5$ amperes.
2. The resistance of L_1 is $R_1 = E_s/I_1 = 100/0.5 = 200$ ohms. Similarly, $R_2 = 400$ ohms, $R_3 = 133$ ohms, $R_4 = 66.7$ ohms, $R_5 = 50$ ohms, and $R_6 = 40$ ohms.
3. The current in branch a is $I_1 + I_2 + I_3 = 0.5 + 0.25 + 0.75 = 1.5$ amperes.
4. The current in branch b is $I_4 + I_5 + I_6 = 1.5 + 2.0 + 2.5 = 6$ amperes.
5. The total circuit current is $I_a + I_b = 1.5 + 6.0 = 7.5$ amperes
6. The total circuit resistance is $R_t = E_t/I_t = 100/7.5 = 13.3$ ohms.
7. The total power supplied to the circuit is $50 + 25 + 75 + 150 + 200 + 250 = 750$ watts. To check the total power in the circuit, note that we can write:

$$P_t = EI_t = 100 \times 7.5 = 750 \text{ watts}$$

LINE DROP IN PARALLEL CIRCUITS

We have learned in our discussion of series circuits that line drop is sometimes excessive in a long line, even though the current-carrying capacity of the wire is not exceeded. Electricians have the same basic problem in various parallel circuit installations. Line drop is of concern when supplying power to electric lamps, because a small reduction in voltage to a lamp results in greatly reduced light output. For example, if a tungsten lamp is operated at 5% less than normal voltage, its light output is 17% less than normal. Therefore, unless the line drop is quite small, the lamps in Fig. 29 will become dimmer as we proceed from the main line down to the end of the branch line. One way to avoid this difficulty is to use large-diameter wire in the branch line.

However, large-diameter wire is comparatively expensive, and we prefer to use as small a diameter as possible as long as the current-carrying capacity of the wire is sufficient. The line drop that causes the lamps to become dimmer as we go down the branch line can be made much less evident by using the *return-loop* system of wiring, as shown in Fig. 30. We can see that the distance of any lamp from

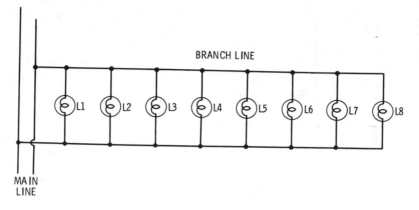

Fig. 29. Variation in lamp brightness due to branch line drop.

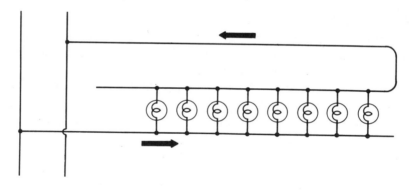

Fig. 30. Return loop system.

the main line is the same in the return-loop system. Therefore, the lamps in this branch circuit have practically the same brightness, particularly in a case where the lamps are spaced equal distances apart. The chief disadvantage of the return-loop system is that it is a three-wire system.

Note, however, that in halls and auditoriums, the lamps are often arranged around a square, rectangle, or circle. Fig. 31 shows a circular arrangement of lamps as used in an auditorium, or on some theater facades. This is called a *re-entrant system,* and it requires only two wires. All of the lamps have practically the same brightness, particularly when the lamps are spaced equal distances apart. There-

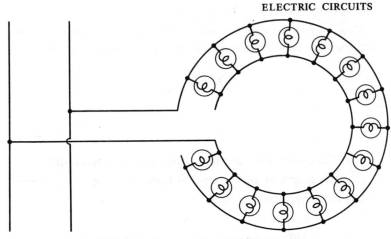

Fig. 31. Re-entrant arrangement of lamps.

fore, the re-entrant system is used in a branch line whenever this is possible. The length of wire used in the re-entrant system is the same as in the simple arrangement of Fig. 29; however, smaller diameter wire can be used in the re-entrant system.

To get a certain chosen amount of voltage drop between the main line (service point) and the middle lamp of a re-entrant system, we use a wire of such size that the total current supplied to the lamps would give the chosen voltage drop over a length $L_1 + L_2$ of wire, where L_1 is the length of wire that carries all of the current, and L_2 is ⅜ of the length of wire in which the current is less than the total amount. Let us take a practical example of a re-entrant row of 200 25-watt tungsten lamps. We will assume that the row is 300 feet long; one end of the row in this example is 60 feet, and the other end of the row is 40 feet from the main line, or service point. Assuming that the voltage at the service point is 115 volts, we will find the size of wire that is necessary to give 110 volts at the middle lamp of the row.

$$I = 200 \times 25/110 = 45.4 \text{ amperes}$$

$$L_1 = 60 + 40 = 100 \text{ feet}$$

$$L_2 = \frac{3}{8} \times 600 = 225 \text{ feet}$$

99

Now, to find the size of wire that is required, we note that the area of the wire in circular mils is given by the following formula:

$$A = \frac{10.8LI}{E_d} \text{ circular mils}$$

where,

A is the cross-sectional area of the wire in circular mils,
L is the length of wire (out and back) in feet,
I is the current in amperes,
E_d is the permissible voltage drop.

Therefore, the size of wire required in this example will be:

$$A = \frac{10.8 \times 325 \times 45.4}{115-110} = 31,870 \text{ circular mils}$$

The nearest gauge is a No. 5 wire, which has a cross-sectional area of 33,100 circular mils. Since the odd-numbered sizes of wire are seldom carried in stock by local supply houses, we would probably use either No. 4 or No. 6 wire in this example.

PARALLEL CONNECTION OF CELLS

Cells may be connected in parallel, as shown in Fig. 32, when the load draws a greater amount of current than can be supplied by a single cell. As a practical example, let us assume that dry cells are to be used to supply 1.5 volts to a load that draws a steady current of ½ ampere. A No. 6 dry cell will supply ⅛ ampere over an extended

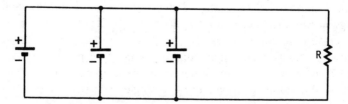

Fig. 32. Circuit arrangement with cells connected in parallel.

time, such as several hours. To meet this requirement, we will connect cells in parallel, as shown in Fig. 33A. In a parallel connection, all positive cell terminals are connected to one side of the line, and all negative cell terminals are connected to the other side of the line. The voltage across the line is the same as the voltage of one cell, or 1.5 volts. However, each cell can contribute its maximum allowable current of ⅛ ampere to the line.

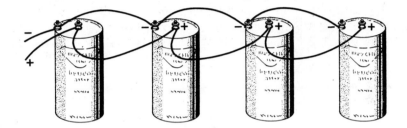

(A) Pictorial view.

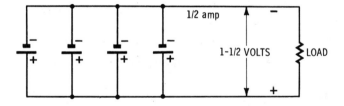

(B) Schematic diagram.

Fig. 33. Dry cells connected in parallel.

Since the load in Fig. 33B draws ½ ampere, we use four cells in this example. Each cell contributes ⅛ ampere to the load. Storage cells may also be connected in parallel when a comparatively heavy current is demanded by a load. If we use lead-acid storage cells, the voltage across the load will be approximately 2 volts. It is interesting to note that a commercial lead-acid storage cell contains several negative plates and an equal number of positive plates connected in parallel, as seen in Fig. 34. When cells are connected in parallel, the result is the same as if we used a single cell with a much larger electrode area.

101

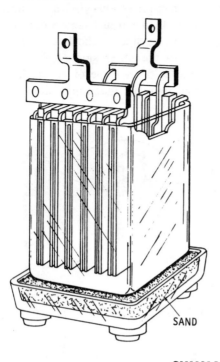

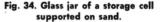

Fig. 34. Glass jar of a storage cell supported on sand.

SAND

SUMMARY

An electric circuit is a closed path for current flow. Current flows from negative to positive, and must be observed when connecting a meter in the circuit. An ammeter is *always* connected in series with a circuit. A voltmeter is *always* connected across a battery, resistor, or any component in the circuit to be measured.

A battery is just one of many components that has a certain amount of resistance, called internal resistance. In many circuits, we can disregard the internal resistance without getting into practical difficulties. Efficiency means output power divided by the input power. For example, if a lamp had no resistance, the efficiency of the circuit would be 100%. We will find that the efficiency of a circuit is greatest when the load is very light—when a very small amount of current is drawn by the load.

In parallel circuits each load is connected in a branch across the voltage source. Therefore, there are as many paths for current flow as there are branches. The loads in a parallel circuit are sometimes said to be connected in multiple. When more loads are connected into a series circuit, the total circuit resistance increases. When more loads are connected into a parallel circuit, the total resistance decreases.

Kirchhoff's law states that just as much electricity must leave a junction as there is electricity entering the junction. This law is important because it helps us to find the amount of branch current that flows in a circuit.

TEST QUESTIONS

1. What is the definition of a series circuit?
2. Explain the meaning of an equivalent circuit.
3. Do resistances add or subtract in a series circuit?
4. What does Kirchhoff's voltage law state?
5. Discuss the meaning of the polarity of a voltmeter. Of an ammeter.
6. How are voltages measured with respect to ground?
7. Why does a dry cell have internal resistance?
8. Explain the difference between the emf and the terminal voltage of a dry cell.
9. How does a battery tester operate?
10. What is the meaning of circuit efficiency?
11. Discuss how we obtain maximum power in a load connected to a voltage source that has internal resistance.
12. If a voltage source is supplying maximum power to a load, what is the circuit efficiency?
13. Give an example of circuit voltages operating in series-opposing.
14. What is the definition of a parallel circuit?
15. How do we find an equivalent circuit that draws the same current as a parallel circuit?
16. Explain two short cuts that can be used to solve certain parallel-circuit problems.
17. What is the meaning of conductance?
18. Do conductances add or subtract in a parallel circuit?
19. How does the light output from a lamp change with a change in voltage?

20. Explain what is meant by a return-loop system.
21. Explain what is meant by a re-entrant system.
22. What is the advantage of a re-entrant system?
23. Why do we connect cells in parallel?
24. How are the positive terminals and the negative terminals of cells connected in a parallel arrangement?
25. Explain why parallel-connected cells operate in the same way as a single cell with a much larger electrode area.

CHAPTER 4

Series-Parallel Circuits

Electricians often work with series-parallel circuits. For example, the circuit shown in Fig. 1A looks like a parallel circuit. If we use a battery that has a very small internal resistance, such as a storage battery, we can say that Fig. 1A is a parallel circuit for all practical purposes. On the other hand, if we use a battery consisting of small penlight cells, the internal resistance of the battery will have to be taken into account. In other words, the complete equivalent circuit is shown in Fig. 1B. We call this a *series-parallel circuit,* because some parts of the circuit are connected in series, and other parts are connected in parallel.

CURRENT FLOW IN A SERIES-PARALLEL CIRCUIT

A series-parallel circuit has branch currents and a total current in the same way that a parallel circuit does. For example, Fig. 2 shows the total current and the branch currents in a simple series-parallel circuit. Let us see how we can find the amounts of these currents in a practical arrangement. If the battery in Fig. 2 supplies 22 volts, and

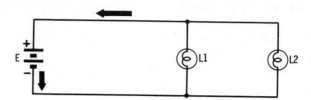

(A) Lamps in parallel connected in series with battery.

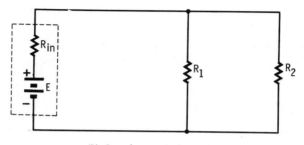

(B) Complete equivalent circuit.

Fig. 1. Series-parallel circuits.

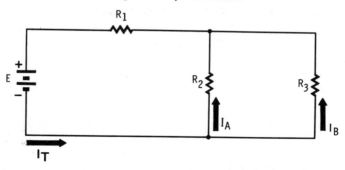

Fig. 2. The total current and branch current.

the resistances of R_1, R_2, and R_3 are 1 ohm, 2 ohms, and 3 ohms, respectively, we reason as follows:

1. R_2 and R_3 are connected in parallel; therefore, their equivalent resistance is 6/5, or 1.2 ohms.
2. R_1 is connected in series with the equivalent resistance of R_2 and R_3; therefore, the total circuit resistance is $1 + 1.2$ ohms, or 2.2 ohms.

106

3. Ohm's law states that $I = E/R$; therefore, I_T is equal to 22/2.2, or 10 amperes.
4. Ohm's law also states that $E = IR$; therefore, the voltage drop across R_1 is equal to 10×1, or 10 volts.
5. The voltage applied to R_2 and R_3 is equal to E minus the voltage drop across R_1; therefore the voltage applied to R_2 and R_3 is 22/2.2, or 10 amperes.
6. Since 12 volts are applied across R_2, I_A is equal to 12/2, or 6 amperes; similarly, I_B is equal to 12/3, or 4 amperes.

It is important to observe that the total current in Fig. 2 is equal to the sum of the branch currents I_A and I_B. That is, we write:

$$I_T = I_A + I_B$$

or,

$$10 \text{ amperes} = 6 + 4 \text{ amperes}$$

KIRCHHOFF'S CURRENT LAW

The foregoing example illustrates an important law of electricity which is called Kirchhoff's current law. This law states: *The sum of the currents leaving a junction is equal to the current entering the junction.* A junction is a circuit point at which a current splits up, or branches, as shown at J in Fig. 3. In this diagram, current I_T is entering junction J, and currents I_A and I_B are leaving junction J. We have seen that I_T is equal to the sum of I_A and I_B.

Kirchhoff's current law is important because it helps us to find the amount of branch current that flows in a circuit such as shown in Fig.

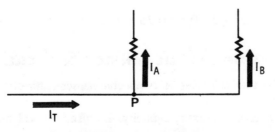

Fig. 3. Illustrating a junction at point P.

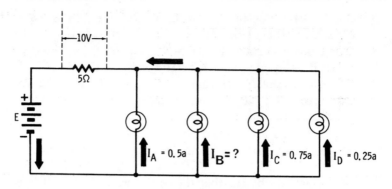

Fig. 4. A series-parallel circuit in which I$_B$ is unknown.

4. In this example, we are given the series resistance and the voltage drop across it, and three of the branch currents; our problem is to find the fourth branch current (I$_B$). We must start by finding the total current. It follows from Ohm's law that we can write:

$$I_T = \frac{10}{5} = 2 \text{ amperes}$$

Since we know the total current, we can use Kirchhoff's current law to find I$_B$:

$$I_A + I_B + I_C + I_D = I_T$$

or,

$$0.5 + I_B + 0.75 + 0.25 = 2$$

Therefore,

$$I_B = 2 - 0.5 - 0.75 - 0.25 = 0.5 \text{ ampere}$$

SERIES-PARALLEL CONNECTION OF CELLS

We know that cells can be connected in series to obtain a greater source voltage; we also know that cells can be connected in parallel to obtain a greater current capacity. In turn, we will find that cells

108

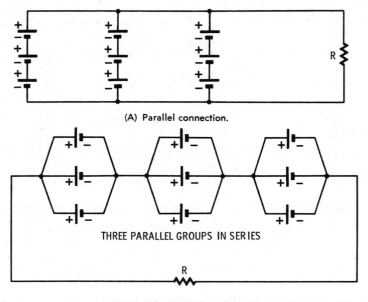

(A) Parallel connection.

THREE PARALLEL GROUPS IN SERIES

R

(B) Series connection.

Fig. 5. Connecting dry-cells in series-parallel.

can be connected in series-parallel to obtain a greater source voltage with a greater current capacity. There are two ways that this can be done, as shown in Fig. 5. We can connect series groups of cells in parallel, or we can connect groups of cells in series. One circuit operates exactly the same as the other; therefore, it makes no difference in practice which method of connection we may use.

LINE DROP IN SERIES-PARALLEL CIRCUITS

The circuit shown in Fig. 6 looks like a parallel circuit, but it is actually a series-parallel circuit. We observe that 117 volts are applied at the input of the line, but there are 114.4 volts at the end of the line. Therefore, the resistance of the 500-foot line is causing a line drop of 2.6 volts. In this example, the wire size for the line is to be found; the wire must have a sufficiently large diameter so that the line drop will not be greater than 2.6 volts. We start by finding the resistance of the line. The lamps draw a total current of 12 amperes. Accordingly, we apply Ohm's law:

109

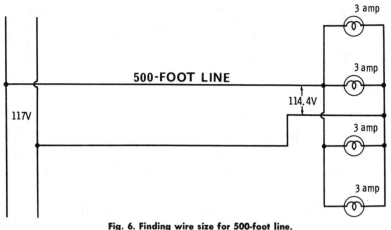

Fig. 6. Finding wire size for 500-foot line.

$$R = \frac{E}{I} = \frac{2.6}{12} = 0.217 \text{ ohm}$$

This is the total resistance of the 500-foot line, which consists of 1000 feet of wire. We know that copper wire has a resistance of 10.4 ohms per circular-mil-foot, or rho is equal to 10.4. Therefore, we write:

$$A = \frac{\rho 2L}{R} = \frac{10.4 \times 2 \times 500}{0.217} = 47{,}926 \text{ circular mils}$$

Since No. 4 wire has a cross-sectional area of 41,700 CM, and No. 3 wire has 52,600 CM, we would select the No. 3 gauge. Finally, we must check on the current capacity of No. 3 wire; since this gauge of rubber-insulated wire is permitted to carry 80 amperes according to the National Electrical Code*, our selection is suitable both from the standpoint of line drop and current-carrying capacity.

---

*Note:

The rules laid down in the National Electrical Code are enforced by means of local ordinances for various cities and towns; these ordinances govern the installation of electric wiring. Note that the detailed requirements set forth in local ordinances may differ to some

USE OF A WATTMETER

We know that power is measured with a wattmeter. A wattmeter combines a voltmeter and an ammeter in one unit, and the scale of the wattmeter reads power values in watts. If we use a voltmeter and an ammeter to measure power, the instruments are connected to a line as shown in Fig. 7A. The ammeter measures the current in the line, and the voltmeter measures the voltage across the load. In turn, we multiply volts by amperes to find the power in watts.

A voltmeter and an ammeter each have two terminals. However, since a wattmeter is a combination voltmeter and ammeter, the wattmeter has four terminals. Two of the wattmeter terminals are voltage terminals, and the other two are current terminals. As shown in Fig. 7B, we connect the voltage terminals *across* the line, and we connect the current terminals in *series* with the line. It is very important not to make a mistake in connecting a wattmeter to a line; if the current terminals were connected across the line by mistake, the wattmeter would be damaged.

CIRCUIT REDUCTION

The reduction of a series-parallel circuit to a simple equivalent circuit is based on the methods used for series circuits and for parallel circuits. For example, reduction of the circuit shown at A-1 in Fig. 8 is very simple if we keep the following facts in mind:

---

extent from the letter of the NEC. Before an electrician starts a job, he should first determine whether local ordinances are in force to specify electric wiring installations. Local ordinances should always be observed; however, if there are no local ordinances, the electrician should follow the requirements laid down in the most recent issue of the National Electrical Code. An installation that is in accordance with the code merely insures minimum risk from fire or accident hazards. It does not provide assurance that the installation will operate satisfactorily or efficiently. In other words, the principles of good practices explained in this book must be observed in order to plan an installation that will operate satisfactorily and efficiently.

111

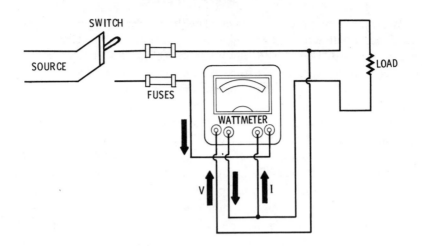

(A) With voltmeter and ammeter.

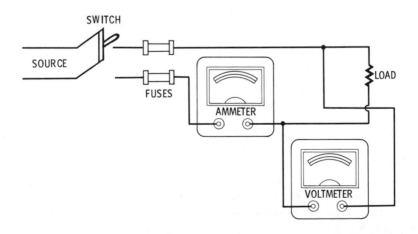

(B) With wattmeter.

Fig. 7. Measuring power.

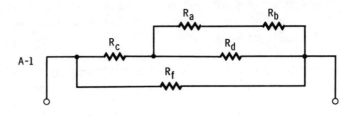

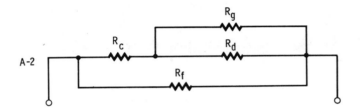

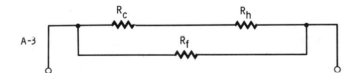

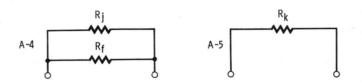

Fig. 8. Reduction of a series-parallel circuit to an equivalent resistance.

113

1. Any number of resistances connected in series can be replaced by a single resistance with a value equal to the sum of the individual resistances.

2. Any number of resistances connected in parallel can be replaced by a single resistance with a value equal to the reciprocal of the sum of the reciprocals of the individual resistances.

For example, if we have three resistances connected in series, we write:

$$R_{eq} = R_1 + R_2 + R_3$$

On the other hand, if we have three resistances connected in parallel, we write:

$$R_{eq} = \frac{1}{\dfrac{1}{R_1} + \dfrac{1}{R_2} + \dfrac{1}{R_3}}$$

We observe that circuit A-1 in Fig. 8 consists of resistors R_a and R_b in series, and that this series combination is in parallel with R_d. In turn, this series-parallel combination is connected in series with R_c, and the resulting combination is connected in parallel with R_f. To find the equivalent circuit, we first replace R_a and R_b with their equivalent resistance R_g, as shown in A-2. The next step is to combine R_g and R_d, replacing them by their equivalent resistance R_h, as seen in A-3. We then Replace R_c and R_h by their equivalent resistance R_j, as depicted in A-4. Finally, we replace R_j and R_f by their equivalent resistance, and obtain the equivalent resistance R_k, as shown in A-5.

Electricians use different ways to reduce resistances in parallel to a single equivalent resistance; the method that is used is simply a matter of personal preference. For example, let us consider the circuit shown in Fig. 9. We can find the equivalent resistance by any of the following methods:

1. *Reduction by Pairs*

In reduction by pairs, we take the resistors two at a time. Thus, we may select R_1 and R_2, and find their equivalent resistance:

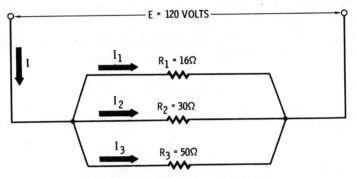

Fig. 9. Reducing resistance in parallel circuits.

$$R_{eq} = \frac{R_1 R_2}{R_1 + R_2} = \frac{16 \times 30}{16 + 30} = \frac{480}{46} = 10.434 \text{ ohms}$$

Next, we will take R_{eq} and R_3, and find their equivalent resistance:

$$R_{eq} = \frac{R_{eq} R_3}{R_{eq} + R_3} = \frac{1200}{139} = \frac{521.7}{60.434} = 8.63 \text{ ohms, approx.}$$

2. *Reduction by Product-and-Sum Formula*

If we prefer, we can take all three resistors at the same time in the product-and-sum formula. Accordingly, we write:

$$R_{eq} = \frac{R_1 R_2 R_3}{R_1 R_2 + R_1 R_3 + R_2 R_3} = \frac{16 \times 30 \times 50}{480 + 800 + 1500} =$$

$$\frac{24,000}{2780} = \frac{1200}{139} = 8.63 \text{ ohms, approx.}$$

3. *Reduction by Reciprocals*

To use the method of reduction by reciprocals, we write the reciprocal of the equivalent resistance as the sum of the reciprocals of the individual resistances:

$$\frac{1}{R_{eq}} = \frac{1}{R_1} + \frac{1}{R_2} + \frac{1}{R_3} = \frac{1}{16} + \frac{1}{30} + \frac{1}{50}$$

or,

$$\frac{1}{R_{eq}} = \frac{75}{1200} + \frac{40}{1200} + \frac{24}{1200} = \frac{139}{1200}$$

$$R_{eq} = \frac{1200}{139} = 8.63 \text{ ohms, approx.}$$

4. *Reduction by Conductances*

In reduction by conductances, we write the equivalent conductance as the sum of the individual conductances:

$$G_{eq} = G_1 + G_2 + G_3 = 0.0625 + 0.0333 + 0.02$$

$$= 0.1158 \text{ mhos, approx.}$$

or,

$$R_{eq} = \frac{1}{G_{eq}} = \frac{1}{0.1158} = 8.63 \text{ ohms, approx.}$$

POWER IN A SERIES-PARALLEL CIRCUIT

The power consumed by the loads in a series-parallel circuit is equal to the sum of the power values consumed by each load. For example, Fig. 10 shows a 3000-watt (3kw) load consisting of a heater, hot plate, and flatiron, each of which consumes 1000 watts. However, the total power in this circuit also includes the power loss to the line due to line drop. The line drop in this example is $117 - 107$, or 10

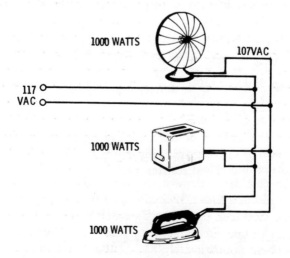

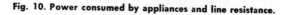

Fig. 10. Power consumed by appliances and line resistance.

volts. To find the power loss in the line, we must first find the current drawn by the load. Accordingly, we write:

$$107I = 3000 \text{ watts}$$

or,

$$I = \frac{3000}{107} = 28 \text{ amperes, approx.}$$

The power loss in the line is equal to the line drop multiplied by the line current:

$$P_{line} = 10 \times 28 = 280 \text{ watts, approx.}$$

Therefore, the power consumption of the circuit shown in Fig. 10 is equal to 3280 watts. The power input to the line is the product of the input voltage and the line current:

$$P_{in} = 117 \times 28 = 3276 \text{ watts, approx.}$$

Note that we find a value of 3276 watts for the power input, but that we found a value of 3280 watts for the circuit power. The reason is that both of these answers are approximate, because we "rounded off" the decimals in our arithmetic.

Next, the *efficiency* of the circuit shown in Fig. 10 is equal to the load power divided by the input power:

$$\text{Efficiency} = \frac{3000}{3276} = 0.91 = 91\%, \text{ approx.}$$

HORSEPOWER

Fig. 11 shows a series-parallel circuit in which the power in the loads is given in horsepower. *One horsepower is equal to 746 watts.* Therefore, the total load power in this example is 746 watts. In turn, the line current will be approximately 7 amperes, and the power input to the line will be approximately 819 watts. The power lost due to line drop is approximately 70 watts. This is a very simple example in

117

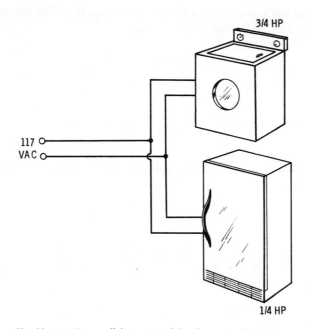

Fig. 11. A series-parallel circuit with loads measured in horsepower.

which we have assumed that the motors in the washer and freezer are 100% efficient. However, no motor is 100% efficient; therefore, a 1-horsepower motor load will actually consume somewhat more than 746 watts from the line. For details of motor power consumption, reference should be made to a specialized motor book.

THREE-WIRE DISTRIBUTION CIRCUIT

A series-parallel circuit of particular importance to electricians is called a three-wire distribution circuit. This type of circuit often operates at a total voltage of 240 volts, while the loads operate at 120 volts. As shown in Fig. 12, a *positive feeder,* a *negative feeder,* and a neutral wire are used in the basic arrangement. The loads are connected between the negative feeder and the neutral, and between the positive feeder and the neutral. When the loads are *unbalanced* (unequal), the neutral wire carries a current equal to the difference of the currents in the positive and negative feeders.

118

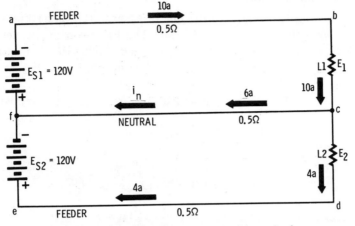

Fig. 12. A three-wire distribution circuit with two loads.

In the example of Fig. 12, load L_1 draws 10 amperes, load L_2 draws 4 amperes, and the neutral wire carries a current of $10-4=6$ amperes. The direction of current flow in the neutral wire is always the same as that of the smaller of the currents in the positive and negative feeders. Note that the current flow is to the left in the positive feeder, and that this current is smaller than the current in the negative feeder. In turn, the current in the neutral wire is in the same direction as in the positive feeder.

With respect to junction c in Fig. 12, we observe that the algebraic sum of the currents entering and leaving the junction is zero:

$$+10-4-6=0$$

Next, let us see how to find the load voltage E_1 in Fig. 12. Since the algebraic sum of the voltages around the circuit $fabcf$ must be zero, we write:

$$+120-10\times0.5-E_1-6\times0.5=0$$

or,

$$E_1=120-5-3=112 \text{ volts}$$

Thus, the voltage across load L_1 is 112 volts. The source voltage is 120 volts, and the line drop is 8 volts. In other words, the line drop in

119

the negative feeder is equal to the line resistance (0.5 ohm) times the line current (10 amperes), or 5 volts. The line drop in the neutral wire is equal to the line resistance (0.5 ohm) times the line current (6 amperes), or 3 volts. Thus, the total line drop between E_{s1} and L_1 is 8 volts.

Load voltage E_2 in Fig. 12 is found in a similar manner. We write:

$$+120 + 6 \times 0.5 - E_2 - 4 \times 0.5 = 0$$

or,

$$E_2 = 120 + 3 - 2 = 121 \text{ volts}$$

In tracing the circuit from f to c, note that we proceed against the direction of the current arrow, and therefore the IR drop of 6×0.5 volts has a plus sign. The load voltage (121 volts) is 1 volt greater than the source voltage (120 volts). The total source voltage is 240 volts, and the total load voltage is $112 + 121 = 233$ volts. Observe that this total load voltage is also equal to the difference between the total source voltage and the sum of the voltage drops in the positive and negative feeders, or, $240 - (2 + 5) = 233$ volts.

When we have a balanced load on the positive and negative sides of a 3-wire system, the current in the neutral is zero, and the currents in the feeders are equal. However, when the loads are unbalanced, the unbalance current flows in the neutral. Therefore, the voltage decreases on the heavily loaded side, and the voltage increases on the lightly loaded side. We see that if the neutral wire had zero resistance, there would be no unbalance in the load voltages. For this reason, it is desirable to use a low-resistance neutral wire when unbalanced loads are present in a system.

A more complicated 3-wire circuit is shown in Fig. 13. The source voltage is 120 volts between each outside wire and the neutral wire. Load currents in the upper side of the system are 10, 4, and 8 amperes respectively, for loads L_1, L_2, and L_3 in this example. In the lower side of the system, the load currents for loads L_4 and L_5 are 12 and 6 amperes, respectively. To find the load voltages, we must find the currents in each outside wire and in the neutral wire. Since the resistances of these wires are given, the voltage drops and the load voltages can be found after the currents are found.

120

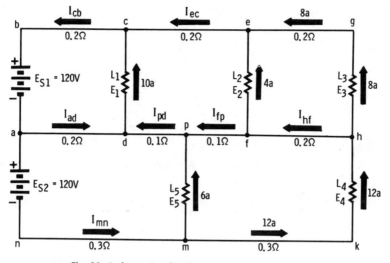

Fig. 13. A three-wire distribution circuit with five loads.

To find the currents, it is best to start at the load farthest from the source in Fig. 13. Electrons flow out of the negative terminal at n and return to the positive terminal at b. It is standard practice to assume that currents flowing toward a junction are positive, and that currents flowing away from a junction are negative. Now, let us apply Kirchhoff's current law at junction h; the neutral current I_n (flowing from h to f) is evidently:

$$12 - 8 - I_{hf} = 0$$

or,

$$I_{hf} = 12 - 8 = 4 \text{ amperes}$$

In the same way, at junctions, $f, e, p, m, d,$ and $c,$ we write:

at f,

$$4 - 4 - I_{fp} = 0$$

or,

$$I_{fp} = 4 - 4 = 0 \text{ amperes}$$

121

at e,

$$4 + 8 - I_{ec} = 0$$

or,

$$I_{ec} = 4 + 8 = 12 \text{ amperes}$$

at p,

$$6 + 0 - I_{pd} = 0$$

or,

$$I_{pd} = 6 + 0 = 6 \text{ amperes}$$

at m,

$$+ I_{mn} - 6 - 12 = 0$$

or,

$$I_{mn} = 6 + 12 = 18 \text{ amperes}$$

at d,

$$I_{ad} + 6 - 10 = 0$$

or,

$$I_{ad} = -6 + 10 = 4 \text{ amperes}$$

at c,

$$I_{cb} + 10 + 12 = 0$$

or,

$$I_c = -10 - 12 = 22 \text{ amperes}$$

Therefore E_{s1} in Fig. 13 supplies 22 amperes and E_{s2} supplies 18 amperes. The electron flow in all parts of the lower wire is outward from the source, and the electron flow in all parts of the upper wire is back toward the source. The current in the neutral wire is always equal to the difference in the currents in the two outside wires, and the electron flow is in the direction of the smaller of these two currents.

In Fig. 13, the neutral current in section ad is 3 amperes, which is the difference between 19 amperes and 22 amperes; also, it is in the

122

direction of the smaller current in section *mn*. The neutral current in
section *pd* is 6 amperes, which is the difference between 18 amperes
and 12 amperes; it is in the same direction as the 12 amperes in section
ec. The neutral current in section *fp* is zero because the current in each
outside wire in that section is 12 amperes. The neutral current in
section *hf* is 4 amperes, which is the difference between 12 amperes
and 8 amperes; it is in the direction of the smaller outside current
in section *ge*.

To find the load voltages in Fig. 13, we apply Kirchhoff's voltage
law to the various individual circuits. Thus, to find the voltage E_1
across L_1, the algebraic sum of the voltages around the circuit *abcda* is
equated to zero. Starting at *a*:

$$-120 + 22 \times 0.2 + E_1 + 4 \times 0.2 = 0$$

or,

$$E_1 = 120 - 4.4 - .8 = 114.8 \text{ volts}$$

To find load voltage E_2, we trace circuit *dcefpd*. Starting at *d*:

$$-114.8 + 12 \times 0.2 + E_2 + 0 \times 0.1 - 6 \times 0.1 = 0$$

or,

$$E_2 = 114.8 - 2.4 - 0 + .6 = 113 \text{ volts}$$

To find voltage E_3, we trace circuit *feghf*. Starting at *f*:

$$-113 + 8 \times 0.2 + E_3 - 4 \times 0.2 = 0$$

or,

$$E_3 = 113 - 1.6 + .8 = 112.2 \text{ volts}$$

To find load voltage E_4, we trace circuit *nadpfhkmn*. Starting at *n*:

$$-120 - 4 \times 0.2 + 6 \times 0.1 + 0 \times 0.1 + 4 \times 0.2 + E_4$$
$$+ 12 \times 0.3 + 18 \times 0.3 = 0$$

or,

$$E_4 = 120 + .8 - .6 - 0 - .8 - 3.6 - 5.4 = 110.4 \text{ volts}$$

To find load voltage E_5, we trace circuit *nadpmn*. Starting at *n*:

$$-120 - 4 \times 0.2 + 6 \times 0.1 + E_5 + 18 \times 0.3 = 0$$

or,

$$E_5 = 120 + .8 - .6 - 5.4 = 114.8 \text{ volts}$$

The foregoing is an example of a comparatively involved series-parallel circuit problem. However, it is a very practical type of problem for the electrician. We recognize that this problem would have been very difficult to solve if we did not understand Kirchhoff's current law and Kirchhoff's voltage law.

SUMMARY

A series-parallel circuit has branch currents and a total current in the same way that a parallel circuit does. In other words, the total current is equal to the sum of the total branch currents.

A wattmeter combines a voltmeter and an ammeter in one unit, and reads power values in watts. If the voltmeter reading and the ammeter reading were multiplied, we would have the power which would be read in watts.

Series-connected cells will produce a greater source voltage, and parallel-connected cells will produce a greater current capacity. In turn, if cells are connected in series-parallel, we can obtain a greater source voltage with a greater current capacity.

Horsepower is a unit of power, or the capacity of a mechanism to do work. It is the equivalent of raising 33,000 pounds one foot in one minute, or 550 pounds one foot in one second. One horsepower is equal to 746 watts.

When we have a balanced load on the positive and negative sides of a 3-wire system, the current in the neutral wire is zero, and the currents in the feeders are equal. When the loads are unbalanced, the unbalanced current flows in the neutral wire. The direction of current flow in the neutral wire is always the same as that of the smaller of the currents in the positive and negative feeders. A 3-wire system often operates at a total voltage of 240 volts, while the load can operate at 120 volts.

TEST QUESTIONS

1. What is the definition of a series-parallel circuit?
2. Explain the meaning of a branch current.
3. State Kirchhoff's current law.
4. What is a junction?
5. Name two ways in which cells can be connected in series-parallel?
6. Why must the branch voltages in a series-parallel arrangement of cells be equal?
7. Discuss how a wattmeter is connected into a circuit.
8. What is the meaning of reduction by pairs? By the product-and-sum formula?
9. What is the meaning of reduction by reciprocals? By conductances?
10. How do we calculate the load power in a series-parallel circuit? The total power?
11. Explain how we find the efficiency of a series-parallel circuit.
12. How many watts are there in one horsepower?
13. What is the basic difference between a three-wire distribution system and a two-wire system?
14. State the arrangement of a neutral wire in a three-wire distribution system.
15. What is the meaning of unbalanced loads?
16. Why do unbalanced loads cause current flow in a neutral wire?
17. Would unbalanced loads have unbalanced voltages if the neutral wire has zero resistance? Why?
18. What is the direction of current in the neutral wire when the loads are unbalanced?
19. Can a load voltage be greater than the source voltage when the loads are unbalanced?
20. Name the laws of electricity that we use in solving a three-wire distribution system.

CHAPTER 5

Electromagnetic Induction

Electricians work with electromagnetic induction in many practical situations. Various electrical devices operate on the principle of electromagnetic induction, and circuits often have a response (circuit action) that is based on this principle. For example, we may note the following typical devices:

Transformers	Generators	Induction motors
Watt-hour meters	Ignition coils	Magnetos
Compensators	Induction regulators	Choke reactors

PRINCIPLE OF ELECTROMAGNETIC INDUCTION

When a permanent magnet is inserted into a coil, as shown in Fig. 1, the voltmeter deflects. In other words, the changing magnetic field in the coil *induces* a voltage in the coil winding. When the permanent magnet is withdrawn from the coil, a voltage of opposite polarity is induced in the coil winding. Note that if the permanent magnet is

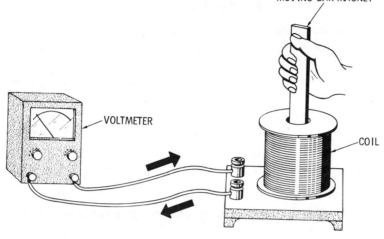

MOVING BAR MAGNET

VOLTMETER

COIL

Fig. 1. An example of electromagnetic induction.

motionless in the coil, there is no induced voltage. Note also that when
the magnet is moved faster, a greater voltage is induced.

It is the *relative* motion of a magnetic field and a wire (or coil)
which induces a voltage in the wire. For example, if a coil is moved
through a magnetic field, a voltage is induced in the coil. This principle
is used in an electrical instrument called a *fluxmeter,* or *gaussmeter,*
to measure the strength of a magnetic field. The instrument consists of
a *flip coil* connected to a *galvanometer,* as shown in Fig. 2. A flip coil
consists of a few turns of wire, usually less than an inch in diameter.
A galvanometer is simply a sensitive current meter. To use a fluxmeter, the electrician places the flip coil in the magnetic field to be
measured, and then "flips" the coil out of the field. In turn, the galvanometer reads the strength of the magnetic field in lines of force
per square inch.

The meter in Fig. 2 is constructed so that the pointer remains at
the point on the scale to which it is deflected when the flip coil is
moved out of the magnetic field that is to be measured. Therefore,
it makes no difference how fast or how slow the electrician moves the
flip coil out of the magnetic field. Before another measurement of
magnetic field strength is made, the pointer is adjusted back to zero
on the scale.

LAWS OF ELECTROMAGNETIC INDUCTION

When a conductor moves with respect to a magnetic field, a voltage is induced in the conductor only if the conductor *cuts* the lines of magnetic force. To cut lines of force means to move at right angles to the lines, as shown in Fig. 3. In this example, conductor *AB* is moving downward through the magnetic field, and this motion is at right angles to the lines of force. The lines of force are directed from the North pole to the South pole of the magnet. A voltage is induced in the conductor. A negative potential appears at *A,* and a positive potential appears at *B.* If we connect a wire from *A* to *B,* electrons flow in the wire as shown.

Electricians need to know the polarity of the voltage induced in a conductor, or, what amounts to the same thing, the direction of induced current in the conductor. This is found by means of *Lenz' law.* Fig. 4 illustrates Lenz' law. If you point in the direction of the magnetic flux with the index finger of your left hand, and point your

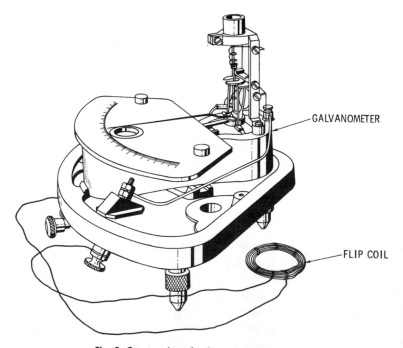

Fig. 2. Construction of a flip coil and galvanometer.

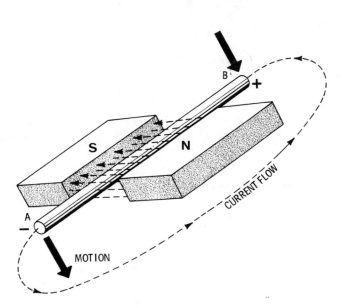

Fig. 3. Induction of voltage in a conductor moving through a magnetic field.

thumb in the direction that the conductor moves, your middle finger then points in the direction of electron flow in the conductor. Note that we hold the index finger, thumb, and middle finger at right angles to one another when we apply Lenz' law.

The amount of voltage that is induced in a moving conductor depends on how fast the conductor cuts lines of magnetic force. If a conductor moves at a speed such that it cuts 10^8 lines of force per second, there will be 1 volt induced in the conductor. Of course, if we use a number of conductors, the induced voltage will be multiplied by the number of conductors. For example, if the coil in Fig. 1 has 100 turns, the induced voltage will be 100 times as great as if only one turn were used.

Since the amount of voltage induced in a conductor depends on how fast the conductor cuts lines of force, it might seem that the

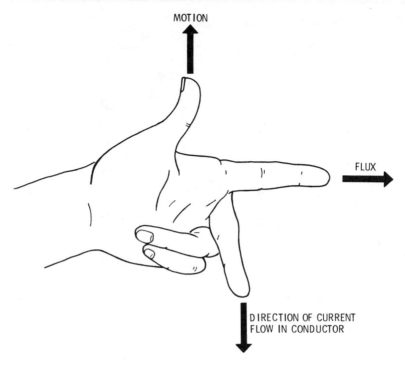

MOTION

FLUX

DIRECTION OF CURRENT
FLOW IN CONDUCTOR

Fig. 4. Illustrating Lenz' law.

reading obtained on a fluxmeter such as shown in Fig. 2 would depend on how fast the flip coil is removed from a magnetic field. However, we will find that the speed with which the flip coil is moved has no effect on the meter reading. The reason that speed makes no difference is because the galvanometer indicates the *total quantity of electricity* that is induced in the flip coil. This total quantity of electricity is the same, whether a small current is induced over a long period of time, or whether a large current is induced over a short period of time. A small current is produced by a small induced voltage, and a large current is produced by a large induced voltage.

SELF-INDUCTION OF A COIL

Electricians often observe that a large flaming spark, or arc, appears across the switch contacts when the circuit to an electromagnet

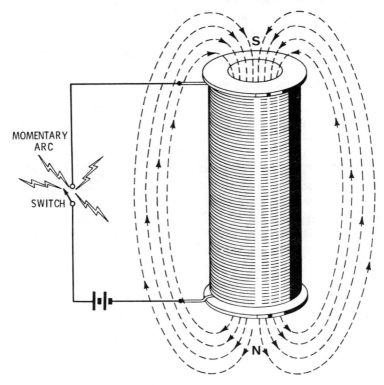

Fig. 5. Self-induction causes arcing at the switch contacts.

is opened. This circuit action is illustrated in Fig. 5. When a larger electromagnet is used, the arc is hotter and larger. This circuit action is explained as follows:

1. When the switch is closed (Fig. 5), current flows through the coil turns and builds up a magnetic field in the space surrounding the coil.

2. A magnetic field contains *magnetic energy;* the larger the coil, the greater is the amount of energy in its magnetic field.

3. At the instant the switch is opened, current flow stops in the the coil, and the magnetic force lines quickly collapse into the coil. This is a moving magnetic field that cuts the coil turns and induces a voltage in the turns—the *voltage of self-induction.*

131

4. While the magnetic flux lines are collapsing, the energy in the magnetic field becomes changed into heat and light energy in the form of a momentary arc between the switch contacts.

There are some important facts concerning self-induction that we should keep in mind. The voltage of self-induction depends on the resistance of the arc path; if the switch is opened quickly, the arc path is longer and the self-induced voltage rises to a higher potential than if the switch is opened slowly. If the switch is opened with extreme rapidity, it may be easier for sparks to jump between turns of the coil than between the switch contacts. The high voltage of self-induction can break down the insulation on the coil wire and damage the magnet. Therefore, electrical equipment that uses large electromagnets carrying heavy current is often provided with a protective device to prevent the voltage of self-induction rising to dangerous potentials.

We will find that the polarity of self-induced voltage is such that it tends to maintain current flow in the same direction as before the switch was opened. The direction of current in the arc between the switch contacts is the same as was flowing in the circuit before the switch was opened. This is often called the *flywheel effect* of an electromagnet. Just as a flywheel continues to turn in the same direction after

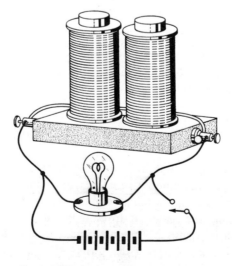

Fig. 6. Self-induction of the electromagnet.

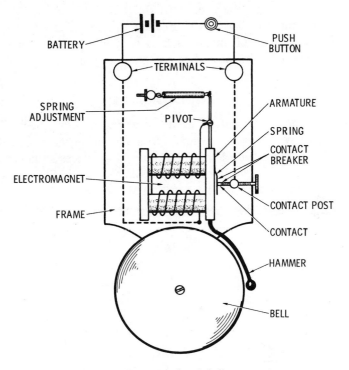

Fig. 7. An electric bell.

an engine is turned off, so does an electromagnet continue to supply current in the same direction to a circuit after the switch is opened.

With reference to Fig. 6, a lamp is connected across an electromagnet. When the switch is closed, the lamp glows dimly, because the coils have comparatively low resistance, and are connected in parallel with the lamp. However, when the switch is opened, the energy in the collapsing magnetic field causes the coil to apply a large momentary voltage across the lamp. Therefore, the lamp flashes brightly. If the coil is sufficiently large, the bulb will burn out when the switch is opened.

An electric bell is illustrated in Fig. 7. The contact breaker automatically opens the circuit to the electromagnet when the armature is attracted; then, the armature springs back and the circuit is again closed through the contacts. This opening and closing action repeats rapidly. Each time that the contacts are opened, a spark jumps between the

133

contacts due to the voltage of self-induction. To prevent the contacts from being burned away in a short time, special metals such as platinum are used. If a soft metal such as copper were used for the contacts, the life of the bell would be comparatively short.

TRANSFORMERS

Transformers are widely used in electrical equipment. In this chapter we will consider certain types of transformers called *ignition coils,* or *spark coils.* Since a bar magnet produces an induced voltage when it is moved into or out of a coil, as shown in Fig. 1, it follows that an electromagnet can be substituted for the bar magnet, as depicted in Fig. 8. It also follows that we can place the electromagnet

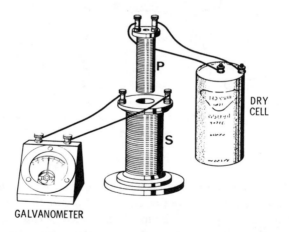

Fig. 8. A moving electromagnet generates a current.

into the coil, and induce a current in the coil by opening and closing the circuit of the electromagnet. *This is an example of transformer action.* A transformer has two windings called the *primary* (P) and the *secondary* (S).

When the primary circuit of a transformer is closed, as shown in Fig. 9A, magnetic flux lines build up in the space surrounding each primary wire *B*. In turn, these expanding flux lines cut each secondary wire *A* and induce a voltage in the secondary. When the primary circuit is opened, the flux lines collapse, and again cut the secondary wire;

134

therefore, a voltage is again induced in the secondary. Note that when the switch is closed, the galvanometer deflects in one direction, but when the switch is opened, the galvanometer deflects in the other direction. The flux lines cut the secondary wire in opposite directions when the switch is opened and closed.

If the switch is held closed in Fig. 9A, the galvanometer does not deflect. In other words, the pointer "jumps" on the galvanometer scale only during the instant that the magnetic field is building up or collapsing. It is easy to see that if the flux lines from a primary wire cut more than one secondary wire, more voltage will be induced in the secondary. (See Fig. 9C.) Since an ignition coil in an automobile must produce a very high voltage in order to jump the gap of the spark plug, the secondary is wound with a large number of turns as compared with the primary. Fig. 10 shows the arrangement of an ignition coil and its circuit. The primary is often called the low-tension winding, and the secondary the high-tension winding.

The ground circuit, shown by dotted lines in Fig. 10, is familiar to every automotive electrician. The engine block and car frame serve as part of the ignition circuit. We observe that both the primary and secondary currents travel through ground circuits. If the metalwork of the car were not used as a ground path for the ignition system, the wiring would be more complicated. Note that a pair of contacts in the distributor are opened and closed by a revolving cam; these contacts operate as a switch to open and close the primary circuit.

A spark coil, or vibrator coil, is similar to an ignition coil except that the contacts are opened and closed by electromagnetic action, as seen in Fig. 11. Note the *capacitor,* which is connected across the vibrator contacts. In actual practice, we will also find a capacitor connected across the contacts in Fig. 10. Let us carefully observe the action of a capacitor, because an electrician is often concerned with this device. A capacitor basically consists of a pair of metal plates spaced near each other, with a sheet of insulating material between the plates, such as waxed paper, mica, or plastic. With reference to Fig. 12, a momentary current flows when the switch is closed. The battery forces electrons into the negative plate, and takes electrons away from the positive plate. As soon as the capacitor *charges* to the same voltage as the battery, current flow stops.

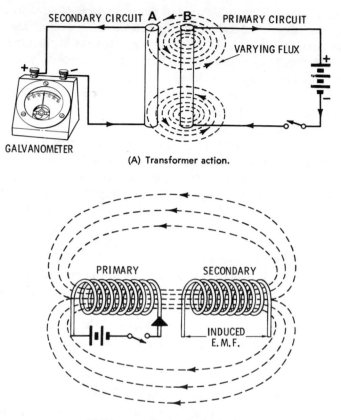

(A) Transformer action.

(B) Primary and secondary coils, showing
magnetic flux lines.

Fig. 9. Basic power

In its charged condition, a capacitor contains electrical energy which can be returned to the circuit. For example, suppose that we open the switch in Fig. 12. The capacitor remains charged because there is no path for escape of electrons from the negative plate to the positive plate. However, if we short-circuit the capacitor terminals (as with a screwdriver), we will observe that a momentary current flows through the short-circuit and produces a snapping spark at the point of contact. The larger the area of the plates in the capacitor, and the closer together the plates are mounted, the greater is the short-circuit on *discharge*.

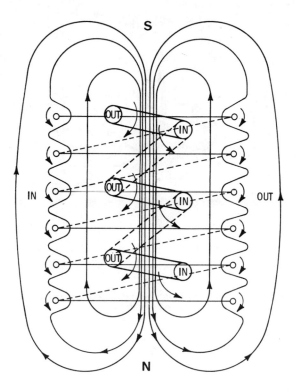

(C) Primary flux lines cutting secondary turns.

transformer action.

Now, let us return briefly to Fig. 11. It is highly desirable to prevent the vibrator contacts from sparking or arcing when the contacts open. The reason for this is because a spark or arc wastes electrical energy that would otherwise appear as induced voltage in the secondary. The capacitor prevents sparking or arcing when the contacts open because the self-induced voltage from the primary is then used to charge the capacitor instead of jumping the air gap between contacts. The self-induced voltage is stored in the capacitor. A short time after the capacitor is charged, it proceeds to discharge back through the primary winding. During the charging process, a little time is provided for the

137

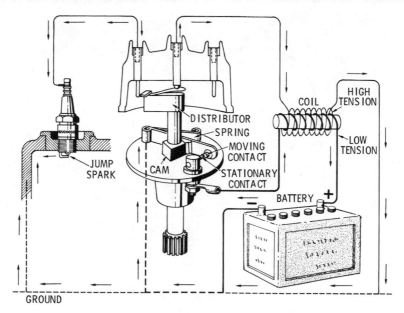

Fig. 10. Diagram of automobile ignition system using battery, ignition coil and high-tension distributor.

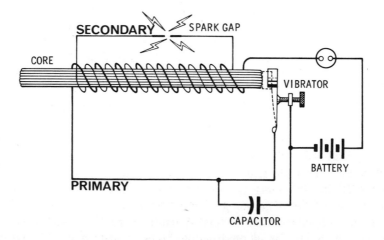

Fig. 11. Diagram of a vibrator spark coil.

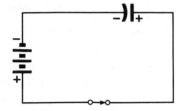

Fig. 12. Capacitor action in a circuit.

vibrator contacts to separate enough that a spark does not jump between them. Therefore, sparking is practically eliminated at the contacts, and the electricity stored in the capacitor next produces a heavy momentary current flow through the primary. In turn, the secondary voltage is much greater than if a capacitor were not used.

Previous mention has been made of electric fences, such as installed by electricians in rural areas. An electric fence operates at sufficiently high voltage that an unpleasant shock is provided upon contact. It usually consists of a single wire supported by insulators on small posts which are spaced considerably farther apart than conventional fence posts. An important feature of the electrical system is the inclusion of sufficient series resistance that current flow is limited to only a few milliamperes. This current limitation guards against the possibility of fatal shock.

An electric fence operates generally at 7500 volts and 2 milliamperes short-circuit current. This is usually a DC voltage that is switched on and off rapidly by a mechanical or equivalent electric device. Interruption of the high voltage insures that an animal or person contacting the electric fence will not "freeze" to the wire and be seriously burned or injured otherwise. The return circuit for an electric fence is provided by a ground connection. Electrical equipment for fence operation is often specified by codes, which may also require that the products have official approval.

Most of these devices are energized by a 117-volt line through an interrupter to the primary of a transformer that resembles a spark coil or ignition coil. The secondary may be wound with sufficiently fine wire that the short-circuit current is automatically limited to a small value. Otherwise, a capacitor or resistor is connected in series with the secondary circuit to limit the short-circuit current as required. In practically all installations, official inspection is required before the system can legally be placed in operation.

ADVANTAGE OF AN IRON CORE

In Figs. 10 and 11, the primary and secondary coils are wound on a single iron core. This core makes the ignition coil or spark coil more efficient than if the primary and secondary were wound on separate iron cores with an air space between them. A single iron core also makes for much greater efficiency than an air core. The reason for this improvement in efficiency is seen in Fig. 13. If an air core or if separate iron cores are used as shown in Fig. 13A and B, most of the magnetic force lines from the primary fail to cut the secondary coil. On the other hand, if a single iron core is used for both coils, as shown in Fig. 13C, most of the primary magnetic force lines then cut the secondary coil.

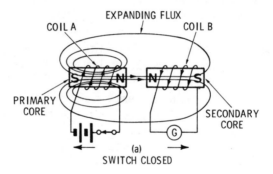

(A) The expanding flux fails to cut the secondary coil.

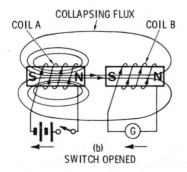

(B) The collapsing flux fails to cut the secondary coil.

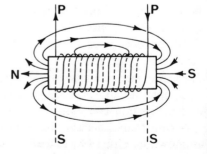

(C) Common core allows most of the primary flux to cut the secondary coil.

Fig. 13. Illustrating single and separate iron cores.

CHOKE COILS

Choke coils are used in various types of electrical equipment. For example, in Fig. 14A, we see how a choke coil is connected in a line conductor with a lightning arrester. If lightning strikes the line, a

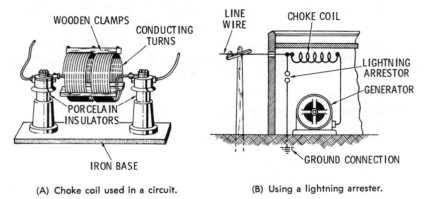

(A) Choke coil used in a circuit.   (B) Using a lightning arrester.

Fig. 14. Illustrating the choke coil and its application.

very large surge of current travels along the line. This heavy current surge would damage equipment at the end of the line (such as the generator shown in Fig. 14B), unless a protective device called a lightning arrester were used. A lightning arrester is basically a spark gap connected between the line and ground.

Since the high voltage produced by a lightning stroke will take the easiest path to ground, the strong electrical surge tends to jump the spark gap to ground instead of flowing through the generator in Fig. 14B. We know that it takes quite a high voltage to break down a spark gap, even when the gap has a comparatively small spacing. Therefore, to make sure that the air gap breaks down before the high-voltage surge reaches the generator, a choke coil is connected between the gap and the generator.

A choke coil consists of a number of turns of large-diameter wire, as shown in Fig. 14A. Since the resistance of the choke coil is very small, the steady current from the generator flows easily through the choke coil, and there is very little voltage drop across the coil. However, when a sudden electrical surge travels down the line, the choke

141

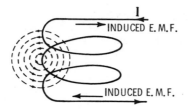

Fig. 15. Illustrating counter electromotive force.

coil blocks the surge and makes it jump the spark gap to ground. Let us see why a choke coil stops a sudden electrical surge.

When a voltage is suddenly applied to a coil, as shown in Fig. 15, the magnetic flux lines build up, or expand. As the flux lines build up, they cut the turns of the coil and induce an emf. Note carefully that *the induced voltage opposes the applied voltage.* This is called inductive opposition to a sudden current change; electricians usually call this opposition *inductive reactance.* Because the sudden current surge is opposed by the induced emf, it takes a certain amount of time for the surge to overcome this opposition and get through the coil.

During the time that the current surge tries to build up a magnetic field in the choke coil, a very high voltage drop is produced across the spark gap, and the surge follows the easiest path to ground by jumping the gap. In other words, before a magnetic field can build up to any great extent in the choke coil, the lightning surge has been harmlessly passed to ground through the spark gap.

Electricians often call the induced emf in Fig. 15 a *counter emf* (cemf), or *back emf.* Counter emf is found in any coil arrangement when voltage is suddenly applied to the coil. Therefore, it takes a certain amount of time for a surge voltage to produce current flow through a coil. On the other hand, after the switch has been closed for a short time in a circuit such as shown in Fig. 16, the current will have risen to its full value given by Ohm's law, and the current thereafter flows steadily as if the coil were not in the circuit.

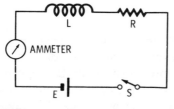

Fig. 16. Illustrating a time delay before
the ammeter indicates a current flow.

REVERSAL OF INDUCED SECONDARY VOLTAGE

We will find that the secondary voltage reverses in polarity when the magnetic flux in a spark coil stops expanding and starts to collapse. For example, we see in Fig. 13A that the expanding lines of force cut the secondary in a direction such that the right end of the secondary core has a South polarity. (This is shown by our left-hand rule.) Next, we see in Fig. 13B that the collapsing lines of force cut the secondary in the opposite direction, and the right-hand end of the secondary core now has a North polarity. Therefore, the induced secondary voltage changes its polarity as the expanding magnetic field changes into a collapsing field.

SWITCHING SURGES

When a switch is closed in a circuit containing a coil, the current rises to a steady value after a short length of time. A magnetic field has then been built up in the space around the coil. Next, when the switch is opened, we know that a momentary arc is produced between the switch contacts, as shown in Fig. 5. Electricians call this self-induced current a *switching surge*. We will now observe how much time it takes the current to build up in a coil circuit such as shown in Fig. 17A when the switch is closed. If a small coil is used, the current builds up fast. On the other hand, if a large coil is used, the current builds up slowly.

The opposition that a coil has to current build-up is measured in *henrys*. Electricians call the *henry* the unit of inductance, just as they call the ohm the unit of resistance. If a coil has an inductance of 1 henry, the current will build up at the rate of 1 ampere each second, as shown in Fig. 17B. In other words, one second after the switch is closed, the current will have built up from zero to 1 ampere. Two seconds after the switch is closed, the current will have built up to 2 amperes, and so on. If there were no resistance in the circuit, and if the battery could supply any amount of current, the circuit current would continue to build up as long as the switch remained closed.

However, we know that even large wire has a small amount of resistance, and that a battery has some internal resistance. Therefore, we cannot have a circuit with zero resistance. This is just another way of

143

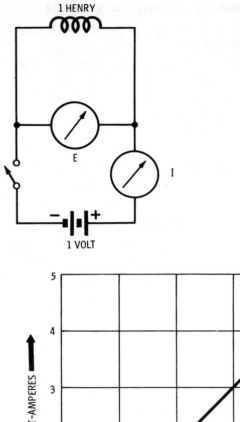

(A) Circuit.

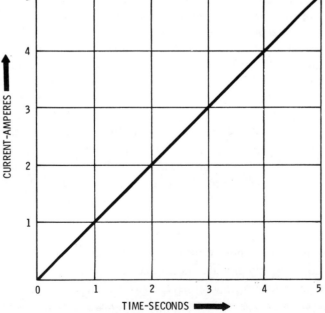

(B) Graph showing current flow versus time.

Fig. 17. A voltage of one volt applied to one henry of pure inductance.

saying that any coil connected to a battery is represented by a circuit containing both inductance and resistance, as shown in Fig. 16. Therefore, the current build-up shown in Fig. 17B will level off after enough time has passed, even if the circuit resistance is very small. This leveled-off value of current is given by Ohm's law: $I = E/R$.

Fig. 18 shows how current builds up in large and small ignition coils. A large coil gives a hotter spark, but the contacts must be allowed to remain closed longer. A small coil gives a weaker spark, but the current build-up levels off quicker. Therefore, to operate a large ignition coil at high speed, we must apply more voltage to the primary so that

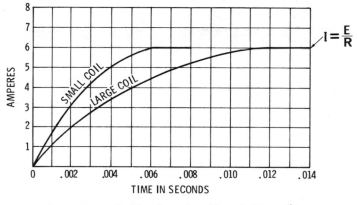

Fig. 18. Current build-up in small and large ignition coils.

the current builds up to a sufficiently high value even though enough time is not available for the current to level off. If an ignition coil has a small primary inductance, it is called a fast coil. On the other hand, an ignition coil that has a large primary inductance is called a slow coil.

THE GENERATOR PRINCIPLE

Various devices that operate on the principle of electromagnetic induction were noted at the beginning of Chapter 5. Although we cannot discuss all of these devices in one chapter, we will conclude with a brief explanation of the generator principle. Generators are familiar to automotive electricians, power-station electricians, and to workers in many other branches of electricity. The basic generator principle was shown in Fig. 3. A voltage is *generated* by induction in a wire that cuts magnetic flux lines.

A generator consists basically of conductors *rotating* in a magnetic field, as shown in Fig. 19. For ease of explanation, a conducting loop is

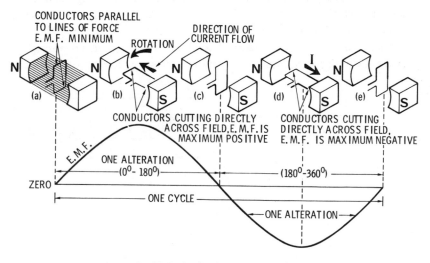

Fig. 19. Basic electric generator action.

depicted that is divided into sections with light and dark lines. In A no lines of force are being cut, and no voltage is induced. However, in B, the loop is cutting the lines of force at right angles, and the induced voltage is greatest. At C, the loop cuts fewer lines of force and the induced voltage decreases. At D, the induced voltage again has its greatest value, but the polarity of the induced voltage is reversed. At E, we have returned to the starting point, and no voltage is induced in the loop at this point in its rotation.

The induced voltage from the basic generator is supplied to an external circuit by means of sliding contacts called *brushes* and *slip rings*. Electricians call the rotating loop assembly an *armature*. The voltage supplied by the generator varies in value periodically and changes its polarity periodically as shown by the curve in Fig. 19. This curve is called a *sine-wave alternating voltage*. If an alternating voltage is applied across a load resistor, a sine-wave alternating current (AC) flows through the resistor. The basic principles of AC current and AC circuit action are explained in the next chapter.

SUMMARY

When a magnet is inserted into a coil that is connected to a voltmeter, the meter will deflect. The changing magnetic field in the coil induces a voltage in the coil winding. It is the relative motion of a magnetic field and a wire which induces a voltage in the wire.

The amount of voltage that is induced in a moving conductor depends on how fast the conductor cuts magnetic lines of force. Since the amount of voltage induced in a conductor depends on how fast the conductor cuts lines of force, it might seem that the reading obtained on a fluxmeter would depend on how fast the flip coil is removed from a magnetic field, but this is not true. The reason that speed makes no difference is because the galvanometer indicates the total quantity of electricity that is induced in the flip coil. A flip coil is a small coil used for measuring a magnetic field. When connected to a galvanometer or other instrument it gives an indication whenever the magnetic field of the coil or its position in the field is suddenly reversed.

Transformers play a very important part in the field of electricity. A transformer is used to convert a current or voltage from one magnitude to another, or from one type to another, by electromagnetic induction. When the primary circuit of a transformer is closed, magnetic flux lines build up in the space surrounding each primary wire. In turn, these expanding flux lines cut each secondary wire and induce a voltage in the secondary. When the primary circuit is opened, the flux lines collapse.

Primary and secondary coils are wound on a single iron core. The system is used in ignition coils and spark coils. A single iron core is more efficient than if the primary and secondary were wound on separate iron cores, because of the air space between the cores.

TEST QUESTIONS

1. What is the principle of electromagnetic induction?
2. Explain Lenz' law.

3. If 1 volt is induced in a moving conductor, how many magnetic lines of force does the conductor cut each second?
4. What is meant by the self-induction of a coil?
5. How can the voltage of self-induction break down the insulation of coil wires when the current to a large coil is suddenly stopped by opening a switch?
6. Discuss transformer action.
7. Why is the secondary of an ignition coil wound with a large number of turns?
8. How can sparking be suppressed between the vibrator contacts of a spark coil?
9. Explain how a capacitor stores electric charge when it is connected to a battery.
10. Why are the primary and secondary of an ignition coil or a spark coil wound on a single iron core?
11. Discuss the polarity of voltage induced in the secondary of an ignition coil when the magnetic flux of the primary is expanding. What happens while the flux is collapsing?
12. What is a choke coil? Give an example of its use.
13. Define counter electromotive force.
14. Why does it require a certain amount of time for a suddenly applied voltage to build up current flow through a coil?
15. If 1 volt is switched across a 1-henry coil, how fast does the current flow build up?
16. What do electricians mean by switching surge?
17. Why does the build-up of current flow through a coil level off to the value given by Ohm's law?
18. Explain the difference between a fast and a slow ignition coil.
19. State what electricians mean by the generator principle.
20. When a conducting loop rotates in a magnetic field, why does the induced voltage rise and fall during a complete revolution?
21. Why does the polarity of induced voltage in an armature loop reverse its polarity each time it completes one-half of a complete revolution?
22. Explain what is meant by a sine-wave alternating voltage.
23. How many cycles are generated by a complete revolution of an armature loop?
24. How many alternations are there in one cycle?

CHAPTER 6

Principles of
Alternating Currents

An alternating voltage or current is usually defined as a voltage or current that changes in strength according to a sine curve. An alternating voltage reverses its polarity on each alternation, and an alternating current reverses its direction of flow on each alternation. Electricians usually speak of "AC voltage" and "AC current." An AC generator is commonly called an *alternator*. Automotive electricians are concerned with the repair and maintenance of alternators. Fig. 1 shows the development of an AC voltage. The point of maximum voltage occurs at 90°, and is also called the *crest* voltage or *peak* voltage of the sine curve.

FREQUENCY

The frequency of an AC voltage or current is its number of cycles per second. For example, electricity supplied by public utility companies in the United States has a frequency of 60 cycles per second (60 cps). Electricians sometimes call cycles-per-second by the newer

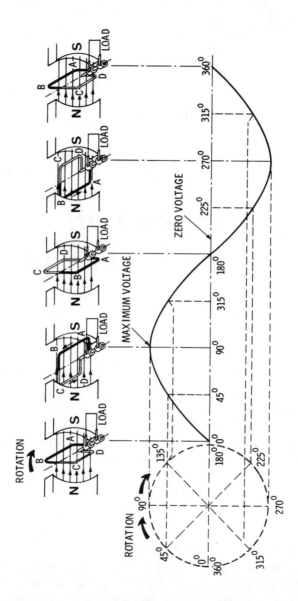

Fig. 1. Simple illustration showing generation of a sine curve during an alternating current cycle.

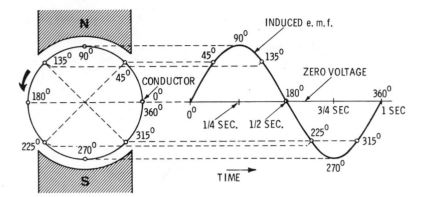

Fig. 2. Curve showing relationship between time and induced voltage in an alternating current circuit.

name *hertz*. Thus, 60 cycles per second is the same as 60 hertz (60 Hz). Fig. 2 shows the relation between time and voltage at a frequency of 1 cps. Each alternation is completed in ½ second. In this example, the time for one complete cycle, called the *period* of the AC voltage, is 1 second. If an AC voltage has a frequency 60 cps, its period is 1/60 second.

INSTANTANEOUS AND EFFECTIVE VOLTAGES

An *instantaneous* voltage is the value of an AC voltage at a particular instant. For example, the maximum voltage shown in Fig. 1 is an instantaneous voltage; the positive maximum voltage occurs at 90°, and the negative maximum voltage occurs at 270°. Electricians generally consider that the most important instantaneous AC voltages are the maximum (crest, or peak) voltage, and the *effective* voltage. We will first explain what is meant by the effective voltage, and then observe why it is so important. The effective voltage is 70.7% of the peak voltage, and occurs at 45°, as shown in Fig. 3.

Practically all AC voltmeters read the effective value of an AC voltage. For example, the AC voltmeter illustrated in Fig. 4 has its scale marked to indicate effective voltages. An effective AC voltage may also be called a root-mean-square (rms) voltage. Since nearly all AC voltmeters indicate effective values, this particular value is always under-

151

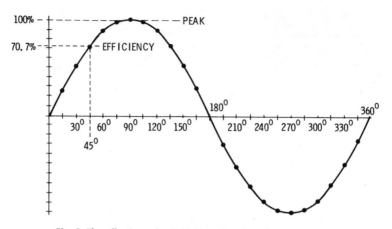

Fig. 3. The effective value is 70.7% of peak and occurs at 45°.

Fig. 4 An AC voltmeter that indicates effective voltage.

stood when an electrician speaks of "117 volts," "234 volts," etc. The word "effective" is understood although it is not commonly stated in electrical diagrams and instruction sheets. If an electrician wishes to speak of a crest or peak voltage, he will state "165 crest volts," or "165 peak volts," for example. An effective voltage of 117 volts has a crest voltage of approximately 165 volts.

Let us see why effective AC volts are so important in practical work. For example, Fig. 5 shows the arrangement of a resistance element in

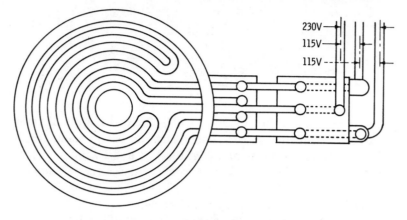

Fig. 5. The resistance element in an electric range.

an electric range. A three-wire distribution system is used, with 230 volts between the feeders and 115 volts between each feeder and the neutral wire. We will find that the resistance element will have the same temperature, whether we use 230 DC volts between the feeders, or use 230 effective AC volts between the feeders. In other words, *a given effective AC voltage provides the same amount of power as the same value of DC voltage.*

As another practical example, let us consider an electric light bulb that is rated for operation on 117 volts AC. This means that the bulb is rated for operation on 117 effective volts of AC. We will find that the bulb provides the same amount of light if it is operated on 117 volts DC. Similarly, a soldering iron, toaster, percolator, or space heater draws the same amount of power from either a 117-volt DC line or a 117 effective-volt AC line. The power in watts consumed by a resistance is equal to E^2/R, whether E is taken in DC volts or in effective AC volts.

OHM'S LAW IN AC CIRCUITS

In most practical electrical work with AC circuits, we use effective values of voltage and current in Ohm's law. Just as an effective voltage is 70.7% of the peak voltage, so is an effective current 70.7% of the peak current. In terms of effective (rms) values of current and voltage, we write Ohm's law:

153

$$I_{eff} = \frac{E_{eff}}{R}$$

or,

$$I_{rms} = \frac{E_{rms}}{R}$$

As we proceed with our discussion of AC circuits, we will simply write I and E with the understanding that the letters stand for I_{eff} and E_{eff}, or, which is the same thing, for I_{rms} and E_{rms}.

Electricians sometimes work with peak voltages (crest voltages). We know that an effective voltage is 0.707 of the peak voltage. Therefore, a peak voltage is 1.414 times the effective voltage. Similarly, a peak current is 1.414 times the effective current. Therefore, we may write Ohm's law for peak values, as follows:

$$1.414\ I_{eff} = \frac{1.414\ E_{eff}}{R}$$

or,

$$I_{peak} = \frac{E_{peak}}{R}$$

Once in a while we work with instantaneous voltages and currents. We write Ohm's law for instantaneous voltages and currents as follows:

where,

$$i = \frac{e}{R}$$

small letters i and e represent instantaneous values.

In other words, Ohm's law is true for effective, peak, or instantaneous values of voltage and current. The only thing that we must watch for is to use the *same kind* of values in Ohm's law. We would get incorrect answers if we used a peak voltage value and an effective current value in Ohm's law. With reference to Fig. 3, various instantaneous values are indicated by dots along the sine curve. We see that

154

since the instantaneous value at 45° is the same as the effective value, Ohm's law for effective values is merely a special case of Ohm's law for instantaneous values. Similarly, since the instantaneous value at 90° is the same as the peak value, Ohm's law for peak values is merely a special case of Ohm's law for instantaneous values.

It would be very tedious to draw sine waves to represent AC voltages. Therefore, electricians represent the armature of a basic AC generator (Fig. 1) by means of an arrow, as shown in Fig. 6. This arrow

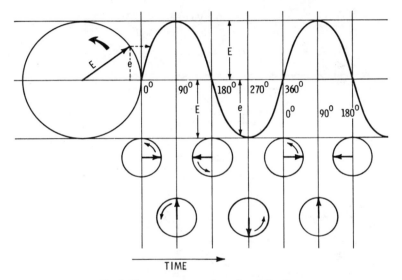

Fig. 6. Vector representation of an AC voltage.

is called a *vector*. As the arrow rotates, it generates a sine curve of voltage, just as a loop of wire rotating in a magnetic field generates a sine curve of voltage. A sine curve is usually called a sine wave because its shape suggests a water wave. Note that the length of the vector E in Fig. 6 represents the peak voltage of the sine wave, and the dotted line which has a length e represents the instantaneous voltage of the sine wave.

When an AC voltage is applied across a resistance, a sine-wave current flows through the resistance, as shown in Fig. 7B. Note that the voltage and current vectors rotate together, and that the current rises and falls "in step" with the voltage. For example, at time t_1, the voltage

155

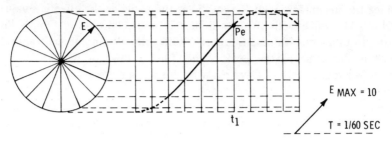

(A) Voltage.

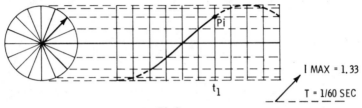

(B) Current.

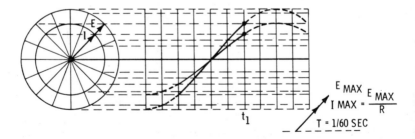

(C) Current and voltage.

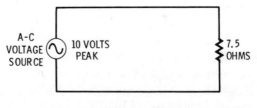

(D) Circuits.

Fig. 7. Vector representation of AC voltage and current.

in Fig. 7A has an instantaneous value indicated by point P_e; at this instant, the current in Fig. 7B has an instantaneous value indicated by point P_1. Fig. 7C shows how the voltage and current vectors rotate together.

In the foregoing example, the peak voltage is 10 volts, which is applied across a resistance of 7.5 ohms, as shown in Fig. 7D. Since $I = E/R$, the peak current is 1.33 amperes. Since the frequency is assumed to be 60 cps (60 Hz), the current and voltage go through their peak positive values each 1/60 second, and go through their peak negative values each 1/60 second. Thus far, we see that Ohm's law is applied to resistive AC circuits in the same way that Ohm's law is applied to resistive DC circuits.

POWER LAWS IN RESISTIVE AC CIRCUITS

Let us consider the power that is consumed in the 7.5-ohm resistor in Fig. 7. Because AC voltage and current have instantaneous, effective, and peak values, it follows that we can find three power values in the 7.5-ohm resistor. The basic power laws are written:

$$\text{Power} = EI = I^2R = \frac{E^2}{R}$$

Let us consider the use of effective voltage and current values in the power laws. This is the most important case, because effective power corresponds to DC power, as we have learned. In other words if we find the power consumed by a resistor when an effective voltage is applied, the same power will be consumed if we apply a DC voltage equal to the effective voltage. In terms of effective voltages and currents, the power laws are written:

$$\text{Effective power} = E_{eff}I_{eff} = I^2_{eff}R = \frac{E^2_{eff}}{R}$$

Therefore, the effective power in the 7.5-ohm resistor (Fig. 7) can be found by multiplying the effective voltage by the effective current. The effective (rms) voltage is equal to 10×0.7071, or 7.071 volts. The effective current is equal to 1.33×0.7071 or 0.94 ampere. Therefore, we write:

157

$$\text{Watts}_{\text{eff}} = 7.071 \times 0.94 = 6.65, \text{ approximately}$$

Of course, the power in the resistor rises and falls. The *peak power* in the resistor is given by:

$$\text{Peak power} = E_{\text{peak}}I_{\text{peak}}$$

Therefore, the peak power in the example of Fig. 7 is found as follows:

$$\text{Watts}_{\text{peak}} = 10 \times 1.33 = 13.3, \text{ approximately}$$

Note that *effective power is equal to ½ of peak power.* Since the power rises and falls in the resistor, we have at any instant an *instantaneous power* value in the resistor:

$$\text{Instantaneous power} = ei$$

or,

$$\text{Watts}_{\text{instantaneous}} = ei$$

or,

$$w = ei$$

Electricians represent instantaneous power by a small letter w, just as e represents instantaneous voltage, and i represents instantaneous current. It is obvious that the instantaneous power value at 45° in Fig. 7 is the same as the effective power value, and that the instantaneous power value at 90° is the same as the peak power value. Note that power does not depend on frequency; the power value is the same, regardless of frequency.

COMBINING AC VOLTAGES

We know how to combine DC voltages by connecting cells in series to form a battery. We also know how to combine voltage drops when we trace around a DC circuit according to Kirchhoff's voltage law. There are certain types of electrical jobs in which an electrician needs to know how to combine AC voltages. To understand how this is done, let us consider the simple example shown in Fig. 8. Two lamps are connected in series with an AC source. The source supplies 234 volts; the filament resistances of the lamps are the same, and there is a voltage drop of 117 volts across each lamp.

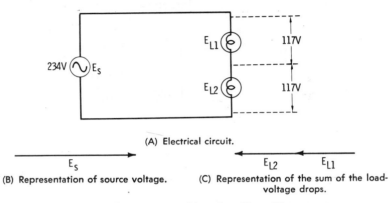

(A) Electrical circuit.

(B) Representation of source voltage.

(C) Representation of the sum of the load-voltage drops.

Fig. 8. Two lamps connected in series with an AC source.

We see that Kirchhoff's voltage is applied to the circuit in Fig. 8 just as if it were a DC circuit. It will be helpful for us to observe how the vector voltages combine. The source voltage E_s is represented by a vector as shown in Fig. 8B. We know that the *polarity of the load* voltage opposes the *polarity of the source* voltage at any given instant. Therefore, the load-voltage vectors are drawn as shown in Fig. 8C. Note that each vector is half the length of the source vector, and each load-voltage vector points in the *opposite* direction to the source vector. We add the load-voltage vectors by placing them end-to-end. Observe that since $E_{L1} + E_{L2}$ has the same length as E_s, but points in the op-posite direction, the vectors cancel; this shows that the algebraic sum of the voltage drops around the circuit is equal to zero.

Note that E_s, E_{L1}, and E_{L2} have the same frequency in Fig. 8. Vectors can be used to represent AC voltages, provided the voltages are sine waves, and provided that they have the same frequency. Vectors can also be used to combine sine-wave voltages that have the same frequency, but which are more or less "out of step." For ex-

159

ample, let us consider the two AC voltage sources shown in Fig. 9. A pair of sine-wave voltage sources E_1 and E_2 are shown in Fig. 9A. We know that cells can be connected in series-aiding or in series-opposing. If the two AC voltage sources are connected so that the sine waves are "in step" as shown in Fig. 9B, the voltages are in series-aiding and the voltages add. On the other hand, if the sine waves are completely "out of step", as shown in Fig. 9C, the voltages are in series-opposing and the voltages subtract.

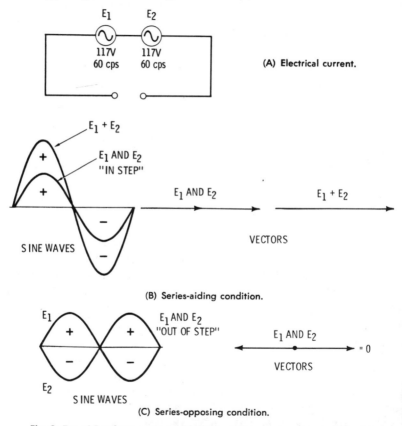

(A) Electrical current.

(B) Series-aiding condition.

(C) Series-opposing condition.

Fig. 9. Two AC voltage sources with the same frequency, connected in series.

The foregoing examples are very simple. However, electricians may have to work with equipment in which a pair of sine-wave voltages that have the same frequency are "half-way out of step," as shown in

160

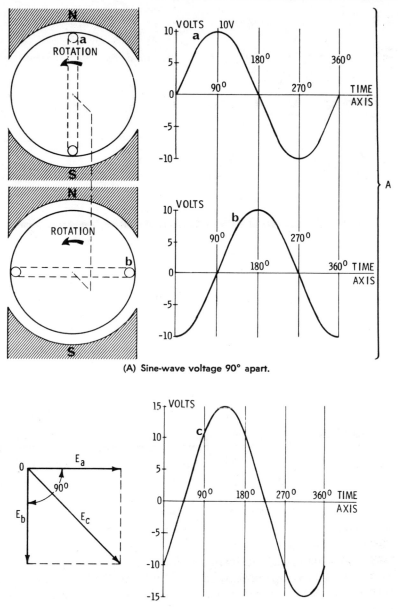

(A) Sine-wave voltage 90° apart.

(B) Vector representation of an AC voltage.

Fig. 10. Two AC voltages of the same frequency that are half-way out of step.

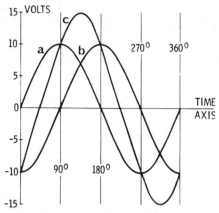

(C) The sum of a and b.

Fig. 10. Two AC voltages of the same frequency are half-way out of step. (cont'd).

Fig. 10. The peaks of the two sine-wave voltages are 90° apart. In this example, the AC voltage sources are represented as a pair of generators; however, we will find that other devices such as capacitors or coils produce sine waves that have peaks 90° apart. It is easier to understand such situations if we start with generators such as shown in Fig. 10.

The sine-wave voltages generated in loops *a* and *b* (Fig. 10A) are 90° apart because the loops are located 90° apart on the two-pole armatures. These armatures are assumed to be mounted on a common shaft in this example. Note that when loop *a* is cutting squarely across the magnetic field, loop *b* is moving parallel to the field and is not cutting magnetic lines of force. Therefore, the voltage in loop *a* is maximum when the voltage in loop *b* is zero. When two voltages are 90° apart, electricians say that the two voltages are 90° *out of phase*. Or, they will say that the *phase angle* between E_a and E_b is 90°, as shown in Fig. 10B.

If loops *a* and *b* in Fig. 10 are connected in series, and the maximum voltage generated in each loop is 10 volts, these voltages do *not* combine to give a maximum voltage of 20 volts simply because the maximum voltages of the loops do not occur at the same time. The peak voltages are separated by 90°, or by ¼ cycle. We cannot merely add the two voltages, because they are out of phase. However, we can add E_a and E_b vectorially, as shown in Fig. 10B, and their vectorial sum is given by E_c.

162

In Fig. 10B, E_c is the vector sum of E_a and E_b; E_c is the diagonal of the parallelogram, the sides of which are E_a and E_b In this example, the parallelogram is a square, because $E_a = E_b$. Therefore, E_c is equal to $\sqrt{2}$ times 10 volts, or approximately 14.14 volts. Let us observe Fig. 10C. This diagram shows how the sine waves E_a and E_b add to give E_c. In other words, if we go along point-by-point and add the instantaneous values of E_a and E_b, we will get the sine wave E_c. We observe that it is much easier and quicker to find the vector sum of E_a and E_b.

Fig. 11 shows how the square on the hypotenuse of a right-angled triangle is equal to the sum of the squares on the other two sides. This is a very important fact for us to remember, because it shows us how two AC voltages are combined when the voltages are not the same. For example, let us suppose that two AC voltages of the same frequency have peak values of 8 volts and 6 volts and are 90° different in

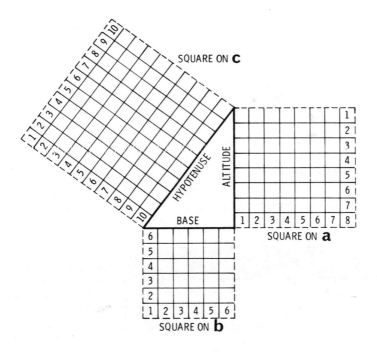

Fig. 11. The square on the hypotenuse is equal to the sum of the squares on the sides of the right-angled triangle.

phase. As shown in Fig. 11, the vector sum of E_a and E_b is 10 volts. If we prefer, we can find E_c by writing:

$$E^2_c = E^2_a + E^2_b$$

or,

$$E_c = \sqrt{E^2_a + E^2_b}$$

In the foregoing example, $E_a = 8$ volts, and $E_b = 6$ volts. Therefore:

$$E_c = \sqrt{8^2 + 6^2} = \sqrt{64 + 36} = \sqrt{100} = 10 \text{ volts}$$

Whether we use vectors or arithmetic, we find that E_c has a peak voltage of 10 volts. To find the effective or rms value of E_c, we multiply 10 by 0.707, to obtain 7.07 volt rms. The principle of vector sums is the most important principle of AC for us to keep in mind, because we will work with vector sums again and again as we proceed with our study of practical electricity.

TRANSFORMER ACTION IN AC CIRCUITS

Transformers are among the most basic and useful devices used in AC circuits. We have previously considered transformer action in

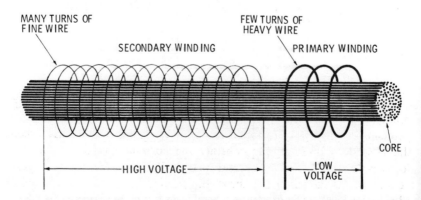

Fig. 12. Transformer consisting of one core and primary and secondary windings.

ignition coils and spark coils. Now, we must see how transformers operate in AC circuits. The arrangement of an elementary transformer is shown in Fig. 12. In this example, the primary has only a few turns while the secondary has many turns. An iron core is used to obtain good efficiency by making most of the primary magnetic flux cut the secondary. If we apply a low AC voltage to the primary of the transformer, we will get a high AC voltage from the secondary, as shown in Fig. 13.

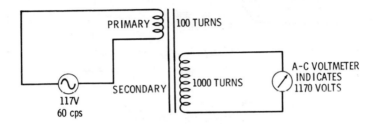

Fig. 13. Step-up transformer action.

When the secondary has more turns than the primary, a transformer is called a *step-up* transformer. The voltage step-up ratio is equal to the turns ratio:

$$\frac{E_{pri}}{E_{sec}} = \frac{N_{pri}}{N_{sec}}$$

where:

E represents AC voltage,
N represents number of turns.

Open-core transformers, as shown in Fig. 14A, are used in some branches of electrical work. However, closed-core transformers as depicted in Fig. 14B are much more widely used, particularly when the secondary must supply considerable current. Closed-core transformers are more efficient because practically all of the primary magnetic flux cuts the secondary turns. Electricians call the construction in Fig. 14B a *shell-type* core.

165

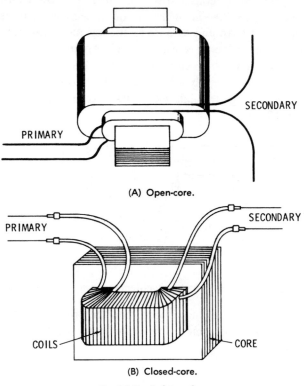

(A) Open-core.

(B) Closed-core.

Fig. 14. Typical transformers.

If a transformer has more turns on the primary than on the secondary, it is called a *step-down* transformer. The transformer shown in Fig. 15 steps down the applied voltage by a 4-to-1 ratio. In this ex-

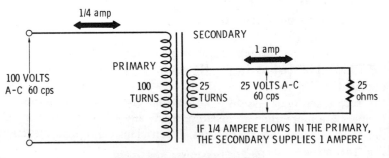

Fig. 15. Step-down type transformer.

ample, a 25-ohm load resistor is connected to the secondary terminals. Note carefully that a step-down transformer steps up the secondary current. On the other hand, a step-up transformer steps down the secondary current. In the example shown in Fig. 15, the primary voltage is stepped down from 100 volts to 25 volts; the primary current is stepped up from ¼ ampere to 1 ampere.

The primary-to-secondary current ratio is equal to the secondary-to-primary turns ratio:

$$\frac{I_{pri}}{I_{sec}} = \frac{N_{sec}}{N_{pri}}$$

This is just another way of saying that the primary power is equal to the secondary power. For example, in Fig. 15, the primary power is 100×0.25, or 25 watts; the secondary power is 25×1, or 25 watts. Of course, if a transformer has poor efficiency, the secondary power will be less than the primary power. For example, let us suppose that an open-core transformer is used which is only 80% efficient. If the primary draws 25 watts, the secondary will supply only 20 watts. Note that shell-type transformers have a high efficiency which may be as great as 99%.

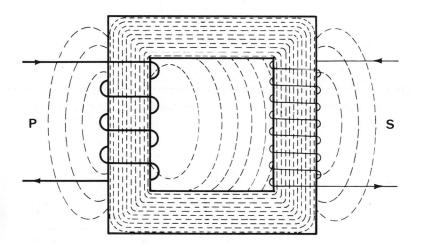

Fig. 16. The core-type transformer.

Another type of closed-core transformer is shown in Fig. 16 and is called a *core-type* transformer. The efficiency of a core-type transformer is between that of an open-core transformer and a shell-type transformer. Note in Fig. 16 that some of the magnetic force lines stray from the iron path into the surrounding air. These stray flux lines are called *leakage flux* and cause reduced efficiency.

Let us consider the AC resistance of the primary and secondary for the loaded transformer shown in Fig. 15. The primary draws 1/4 ampere with 100 volts applied; it follows from Ohm's law that the AC resistance of the primary is equal to E/I, or 400 ohms. The secondary supplies 1 ampere at 25 volts; thus, the AC resistance is 25 ohms. This is an AC resistance ratio of 400/25, or 16-to-1. In other words, the AC resistance ratio of a transformer is equal to the *square* of the turns ratio:

$$\frac{N^2_{pri}}{N^2_{sec}} = \frac{R_{ACpri}}{R_{ACsec}}$$

We have 100 primary turns and 25 secondary turns in Fig. 15. We write:

$$\frac{100^2}{25^2} = \frac{10,000}{625} = \frac{16}{1}$$

Electricians usually call the AC resistance ratio of a transformer its *impedance ratio*. Impedance simply means AC resistance, as will be explained in greater detail later. At this time we are chiefly concerned with the importance of transformer impedance ratio in practical electrical work. For example, we have an AC source in Fig. 17 with an internal resistance of 400 ohms. We will ask how maximum power can be transferred from this source to a 25-ohm load. Maximum power will be transferred if R_L is caused to "look like" it has the same resistance as R_{in}. This is accomplished by the transformer.

Just as we found in Fig. 15, the primary impedance is 400 ohms and the secondary impedance is 25 ohms in this example. Therefore, the source in Fig. 17 works into a 400-ohm primary and maximum power is transferred into the primary. The load works out of a 25-ohm secondary and maximum power is transferred into R_L. There-

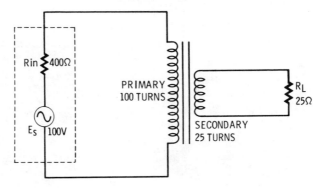

Fig. 17. The transformer turns ratio provides maximum power transfer from the source to the load.

fore, a 4-to-1 turns ratio on the transformer provides maximum power transfer from the source to the load in this example. For an ideal transformer, the system efficiency would be 50%, because half of the available power in the primary circuit is lost in the internal resistance. In turn, if the transformer is 99% efficient, the system efficiency will be 49.5%.

It is important to remember that voltage, current, and power relations in a transformer do not depend on frequency. We could use a 60-cps source or a 1000-cps source in Fig. 17, and the relations of voltage, current, and power would not be changed. However, a core that is very efficient at 60 cps might be less efficient at 1000 cps. We will merely observe in this chapter that the electrician always selects a transformer which is rated properly for the particular application. For example, the transformer in Fig. 15 would be rated for a primary voltage of 100 volts, for 25 watts of power, and for an operating frequency of 60 cps. A 4-to-1 turns ratio would be specified in this application.

In various types of electrical equipment, a transformer may be wound with several secondaries to supply several loads at different voltages. Fig. 18 illustrates small transformers that have several secondary windings. The total power supplied by such a transformer is equal to the sum of the power values supplied by the secondaries. A transformer should never be operated above its power rating. However, a transformer operates satisfactorily at any power value below its power rating. If a mistake is made, and a transformer is operated considerably

169

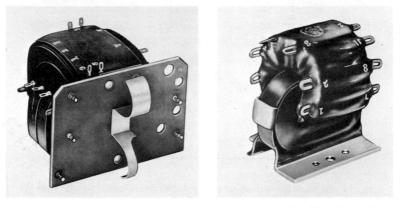

Fig. 18. Typical transformers with more than one secondary winding.

above its power rating, the efficiency will be poor and the transformer will run hot. Insulation is likely to become burned with resulting short-circuits between turns in the coils.

CORE LOSS AND CORE LAMINATION

Iron is the only suitable substance for use in transformer cores because it has a low reluctance. However, iron is also a conductor of electricity. Therefore, AC current in a transformer winding induces voltage in the core, and causes current to flow in the core as shown in Fig. 19. These core currents are called *eddy currents*. Eddy currents are undesirable because they reduce the efficiency of the transformer. In other words, iron has resistance, and eddy currents produce an I^2R core loss. Power that is lost in the core cannot be transferred to the secondary winding.

If iron were an insulator, it would have no eddy-current loss. Therefore, we make the iron core "look like" it has more resistance to eddy currents. This is done by *laminating* the core, as shown in Figs. 19B and C. Note that the path for eddy current flow is longer in B than in A. The path for eddy-current flow is still longer in C. When the total path for eddy-current flow is very long, its resistance is very high; in turn, the amount of eddy current that flows is small.

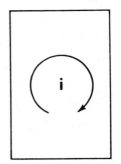

(A) Eddy current in a solid core.

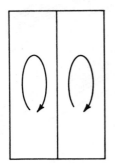

(B) Path of eddy currents increased by core lamination.

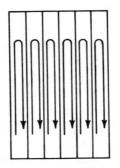

(C) Eddy-current path greatly lengthened by numerous laminations.

Fig. 19. Transformer core action.

We can make the eddy-current value as low as may be desired by simply using more and thinner laminations. Note that laminations must be insulated from one another; otherwise, a laminated core would be the same as a solid core insofar as eddy-current flow is concerned. Laminations are usually lacquered to provide insulation. A transformer core is made by stacking laminations as shown in Fig. 20. Ignition coils and spark coils generally use cores made from bundles of iron wires, as seen in Fig. 12. The iron wires are lacquered to provide insulation and eddy currents are greatly reduced, just as in a laminated core used in transformers.

171

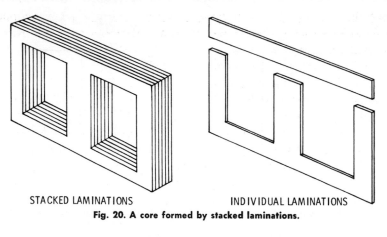

STACKED LAMINATIONS   INDIVIDUAL LAMINATIONS
Fig. 20. A core formed by stacked laminations.

DC VS. AC RESISTANCE

The primary and secondary windings of an ideal transformer would have zero resistance. However, copper wire has a certain amount of resistance. Therefore, the primary and secondary windings have certain DC resistance values. The *winding resistance* of a coil can be measured with an ohmmeter. This DC resistance is undesirable, because it produces an I^2R loss in transformer operation, and reduces the efficiency of the transformer. The I^2R power loss due to the DC winding resistance of the primary and secondary coils is called the *copper loss* of the transformer. This copper loss can be reduced by winding the primary and secondary coils with larger wire. Of course, the DC resistance of the windings cannot be reduced to zero, although it might be so small that it can be neglected in practice.

Let us consider the AC resistance of the primary and secondary. When the secondary of an ideal transformer is open-circuited, the primary would have infinite reactance, and would draw no current from an AC source. However, a practical transformer does not have infinite primary reactance, although this reactance is very high in the case of an efficient transformer. Therefore, the primary of an efficient transformer draws a slight amount of AC current from the source when the secondary is open-circuited. This is called the no-load current. The no-load current is not quite 90° out of phase with the primary voltage, because a practical transformer has copper loss and core loss. These losses consume a small amount of power from the source.

It follows from previous discussion that if a resistance load is connected across the secondary terminals of a transformer, a substantial AC current is drawn by the primary. We say that the connection of the secondary load has caused the primary to have a lower value of AC resistance. Note that this AC resistance cannot be measured with an ohmmeter—the value of the AC resistance is simply a voltage/current ratio. It is the ratio of primary AC voltage to in-phase primary AC current. Even an ideal transformer would have a value of primary AC resistance that depends on the value of the load connected across the secondary terminals.

SUMMARY

Voltage is continually varying in value and reversing its direction at regular intervals in an alternating-current circuit. The frequency of an AC voltage or current is its number of cycles per second, sometimes called *hertz* which is the newer name. An instantaneous voltage is the value of an AC voltage at a particular instant.

The most important instantaneous AC voltages are the maximum (crest, or peak), and the effective voltages. The effective voltage is 70.7% of the peak voltage. The effective voltage is also called rms (root-mean-square) voltage. Peak or crest voltage is the maximum value present in a varying or alternating voltage, and can be either positive or negative.

Transformers are a useful device in AC circuits. If a secondary has more turns than the primary, it is called a step-up transformer. Open-core transformers are used in some branches of electrical work, but closed-core transformers are more efficient because the primary magnetic flux lines cut the secondary supplying considerably more current.

The efficiency of a core-type transformer is between that of an open-core and a shell-type transformer. Some of the magnetic lines of force stray from the core into the surrounding air. These stray flux lines are called *leakage flux* and cause reduced efficiency in the transformer output.

Eddy currents is current developed in the core which reduce the transformer voltage. Since iron is used in transformer cores, and the iron has resistance, the cores are laminated to reduce the path for eddy currents to flow.

TEST QUESTIONS

1. What is an alternator?
2. Explain what is meant by the frequency of an AC voltage.
3. How is the frequency of an AC voltage related to its period?
4. What do we mean by an instantaneous voltage? An effective voltage?
5. Discuss the difference between an effective voltage and a peak voltage.
6. Does Ohm's law hold true for instantaneous, effective, and peak values of AC voltages and currents?
7. What is a vector? Why is it helpful to represent AC voltages and currents by vectors?
8. Explain the difference between peak power, effective power, and instantaneous power.
9. Draw voltage vectors to illustrate Kirchhoffs' voltage law for an AC series circuit.
10. Show how AC voltage sources can be connected in series-aiding and in series-opposing.
11. What is meant by AC voltages that are 90° out of phase?
12. If two AC voltages are equal, but are 90° out of phase, what is their total voltage?
13. If two AC voltages have values of 8 volts and 6 volts, and are 90° out of phase, what is their total voltage?
14. Define a step-up transformer. A step-down transformer.
15. Why does a step-up transformer step the AC voltage up and step the AC current down? (Hint: the primary power is equal to the secondary power.)
16. Define an open-core transformer. A closed-core transformer.
17. What is the difference between a core-type and a shell-type transformer?
18. How is a transformer used to obtain maximum power transfer between an AC source and a load?

19. State the percentage efficiency of a high-quality transformer.
20. Why does stray flux (leakage flux) reduce the efficiency of a transformer?
21. Discuss eddy currents in transformer cores.
22. What is meant by a laminated core?
23. Why must core laminations be insulated from one another?
24. Explain why laminations reduce core loss.
25. Why should a transformer be operated within its power rating?

CHAPTER 7

Inductive and Capacitive AC Circuits

Inductive AC circuits are used in many types of electrical equipment. For example, electricians who work in theaters and motion-picture studios are concerned with lighting control equipment. The basic principle of lighting control is shown in Fig. 1. To control the brightness

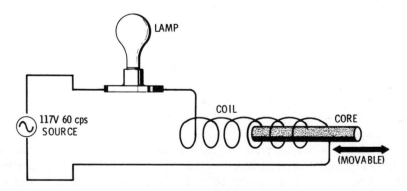

Fig. 1. Basic lighting-control circuit.

176

of the lamp, it is connected in series with a coil having adjustable inductance. The coil has least inductance when the iron core is withdrawn; the coil has greatest inductance when the core is fully inserted. In turn, the circuit current is greatest when the inductance value is least, and the current is small when the inductance is large.

A coil with an adjustable core is called a *variable reactor* or a *variable inductor*. Inductance opposes the flow of AC current because self-induction causes the inductance to have *reactance*. We know that inductance is measured in henrys. We will find that although reactance is different from resistance, reactance is nevertheless measured in ohms. Obviously, electricians must know the difference between reactance and resistance and must know how to find the number of ohms of reactance in a circuit. Therefore, let us consider the relation of henrys to ohms of reactance.

INDUCTIVE CIRCUIT ACTION

We know that inductance opposes any change in current flow. If we increase the voltage across an inductor, an increase in current flow builds up gradually due to the self-induced voltage that opposes current change. On the other hand, if we decrease the voltage across an inductor, the current flow falls gradually due to the self-induced voltage. In other words, the most noticeable difference between a resistive circuit and an inductive circuit is in their response speeds. The current flow in a resistive circuit changes immediately if the applied voltage is changed. On the other hand, the current flow in an inductive circuit is delayed with respect to a change in applied voltage.

We know that inductance is the result of magnetic force lines cutting the coil turns when the magnetic field strength changes. Therefore, the inductance of an air-core coil is greatly increased if we insert an iron core (Fig. 1). Note that if the coil wire is doubled back on itself, as shown in Fig. 2, the magnetic fields will cancel, and the inductance will be zero. This is called a noninductive coil, and is used in electrical equipment in which we wish to employ wire-wound resistors that have no inductance.

In the previous chapter we learned that the unit of inductance is the henry, and that an inductor has an inductance of 1 henry if the current flow increases at the rate of 1 ampere per second when 1 volt is

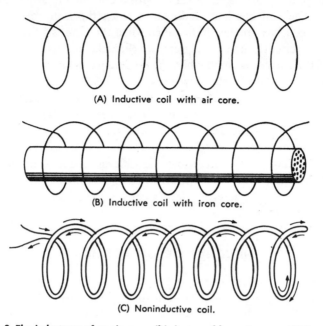

(A) Inductive coil with air core.

(B) Inductive coil with iron core.

(C) Noninductive coil.

Fig. 2. The inductance of an air-core coil is increased by an iron core; if the wire is doubled back on itself, the inductance is zero.

applied across the inductor. We will find that the inductance of a coil increases as the square of the number of turns. For example, with reference to Fig. 3, a coil with two layers of wire has four times as much inductance as a coil with one layer; a coil with three layers of wire has nine times as much inductance as a coil with one layer. Let us see why this is so.

If we wind a single-layer coil, we have a certain number of turns that are cut by a magnetic field of a certain strength. If we wind two layers on the coil, we have twice as many turns that are cut by a magnetic field that is twice as great. Therefore, the self-induced voltage is four times as great as in a single-layer coil. This is just another way of saying that the inductance of a coil increases very rapidly as we add more layers to its winding. We will now assume that we have constructed a 1-henry coil, and we will ask how much opposition it has to AC current flow. To simplify our question, we will assume that the coil has been wound with large wire so that its DC resistance can be neglected.

178

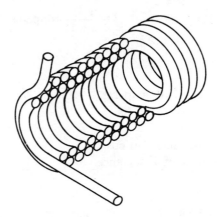

(A) Two-layer.

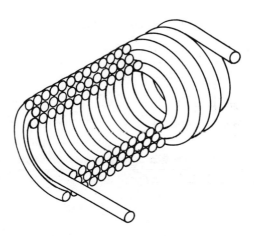

(B) Three-layer.

Fig. 3. Illustrating two and three layer coils.

Electrical measurements with an AC voltmeter and ammeter will show that Ohm's law for an inductor is slightly more complicated than Ohm's law for DC with a resistor. We will find that Ohm's law for an inductor is written:

179

$$I = \frac{E}{2\,\pi\,fL} \text{amperes}$$

where:

I represents AC amperes,
E represents AC volts,
π is 3.1416,
f is the frequency in cps,
L is henrys of inductance.

Therefore, if we apply an AC voltage of 117 volts across an inductance of 1 henry, and the frequency is 60 cps, as shown in Fig. 4, the AC current flow is approximately 0.31 ampere. It is helpful to write Ohm's law for an inductive circuit as follows:

$$\text{AC Current} = \frac{\text{AC Voltage}}{\text{Reactance}}$$

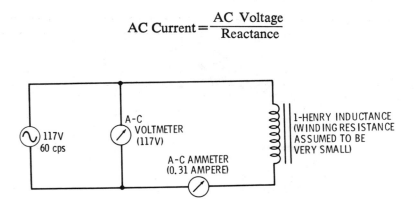

Fig. 4. Illustration of Ohm's law for a purely inductive circuit.

We observe that reactance can be compared with resistance in Ohm's law. In other words, $2\,\pi\,fL$ is measured in AC ohms. Electricians use X_L to represent AC ohms. Thus, $X_L = 2\,\pi\,fL$. Therefore, Ohm's law for inductive circuits is written.

$$I = \frac{E}{X_L}$$

It is obvious that if we used a 117-volt DC source in Fig. 4 instead of a 117-volt AC source, an extremely large current would flow because the wire resistance of the 1-henry coil was assumed to be very small.

Next, let us note how the current varies in the circuit of Fig. 4 when the frequency is changed. Ohm's law for AC in an inductive circuit tells us that if we increase the frequency from 60 cps to 120 cps, the current will be reduced one-half. On the other hand, if we decrease the frequency from 60 cps to 30 cps, the current will be doubled.

If we keep the frequency at 60 cps, but double the inductance to 2 henrys, the current will be reduced one-half. On the other hand, if we reduce the inductance to ½ henry, the current will be doubled. Of course if we keep the frequency and the inductance the same, the current will double if we double the applied AC voltage.

POWER IN AN INDUCTIVE CIRCUIT

We know that the power in a DC resistive circuit is equal to volts multiplied by amperes. This is also true in an AC resistive circuit, as shown in Fig. 5. On the other hand, in a circuit that has inductance

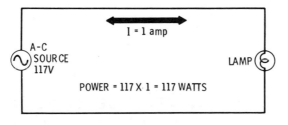

Fig. 5. AC power in a resistive circuit.

only as shown in Fig. 6, the *true power* is zero. This fact might seem puzzling until we observe the circuit action. We recognize that the current flow through an inductor is delayed when voltage is applied due to the self-induced voltage of the inductor which opposes the applied voltage. It can be shown that the AC current in an inductor is delayed 90° with respect to the applied voltage, as seen in Fig. 6.

Now, if we multiply instantaneous voltages by instantaneous currents, we will find the instantaneous power. Over a complete cycle we

181

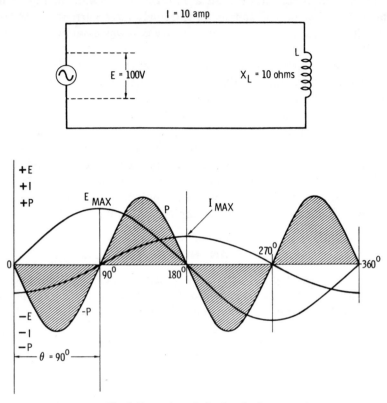

Fig. 6. Power in an inductive circuit.

see that half of the power is below the axis (negative power) and the other half of the power is above the axis (positive power). Therefore, the average power in an inductor is half positive and half negative; no power is consumed by the inductor. So we can simply say that an inductor takes power from the source during the positive alternation and returns this power to the source during the negative alternation.

However, although no power is taken on the average by the inductor from the source in Fig. 6, we see that power is surging back and forth in the circuit—the inductor first stores power in its magnetic field and then returns this stored power to the circuit. Thus, the circuit draws current but *consumes no power*. This is just another way of saying that the source does no work in this circuit, or that the inductor does not actually load the source. The inductor is an apparent load. Current that flows into the inductor on the positive alternation is stored, and this current is then returned to the circuit on the negative alternation.

The product of voltage and current in a purely inductive circuit is called *apparent power* because it is "floating power" that does no work. Apparent power is also called *reactive power,* and is measured in *vars.* We write vars to represent "volt-amperes reactive." In other words, we write, for Fig. 6:

$$\text{Vars} = E_{rms}I_{rms}$$

Electricians measure vars with a varmeter, as shown in Fig. 7. A varmeter looks like a wattmeter, but it is constructed so that it indicates

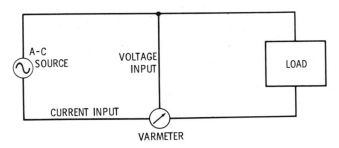

Fig. 7. Measurement of apparent power in an AC circuit.

apparent power instead of true power. Note that if the load were a resistance, such as a lamp, the varmeter would read zero. On the other hand, if the load were a pure inductance, the varmeter would read the product of the rms voltage and the rms current.

RESISTANCE IN AN AC CIRCUIT

We know that in a circuit containing only resistance (Fig. 8), the current and voltage are in phase. Therefore, there is true power in the resistor which appears as heat. True power is also called real power. We also know that this true power in watts is equal to $E_{rms}I_{rms}$. The shaded area in Fig. 8 represents the product of instantaneous voltages and instantaneous currents. The entire shaded area is positive and,

183

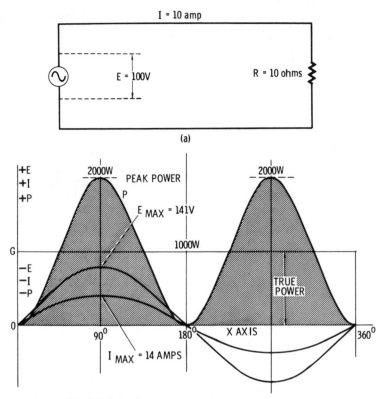

Fig. 8. Relation between E, I and P in a resistive circuit.

accordingly, represents true power. Note that when the voltage is negative, the current is also negative, and therefore their product must be positive.

Next, we will ask what the AC circuit action will be when inductance and resistance are connected in series, as shown in Fig. 9. The AC current I_o flows through both the inductor and the resistor. Accordingly, a voltage drop is produced across the inductor and across the resistor. We observe that the resistive voltage drop E_R is in phase with the current, but that the inductive voltage drop E_L is 90° out of phase with the current. E_R is 50 volts and E_L is 86.6 volts in this example.

Note that the applied voltage E_o in Fig. 9 is the hypotenuse of a right triangle, the sides of which are 50 and 86.6; in turn, we write:

184

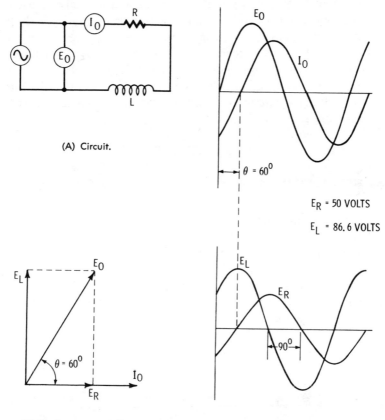

(A) Circuit.

E_R = 50 VOLTS

E_L = 86.6 VOLTS

(B) Vector representation.

(C) Sine-wave.

Fig. 9. Resistance and inductive reactance in series.

$$E_o = \sqrt{50^2 + 86.6^2} = 100 \text{ volts}$$

If we use a protractor in Fig. 9B, we will find that E_o makes an angle of 60° with I_o. Electricians call this angle the *power-factor angle* of the circuit. It is evident that if we keep L constant, and increase R, the power-factor angle will become less. In large factories and power plants, electricians are concerned with power-factor angles. However, details of this kind of electrical work will be explained later.

IMPEDANCE IN AC CIRCUITS

The circuit in Fig. 9 has both resistance and inductive reactance. The total opposition to AC current flow is called *impedance*. Impedance is measured in ohms, just as resistance and reactance are measured in ohms. If an AC circuit contains reactance only, the circuit impedance is the same as the circuit reactance. Or, if an AC circuit contains resistance only, the circuit impedance is the same as the circuit resistance. When a circuit has both resistance and reactance, we find the circuit impedance as follows:

$$Z = \sqrt{R^2 + X^2}$$

where:

Z represents the ohms of impedance,
R represents the ohms of resistance,
X represents the ohms of reactance.

We can draw an impedance diagram as a right triangle, such as shown in Fig. 10. The power-factor angle is the angle *theta* between R and Z in the diagram.

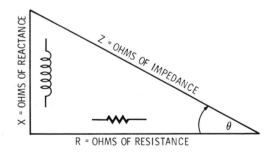

Fig. 10. An impedance diagram.

POWER IN AN IMPEDANCE

When inductance and resistance are connected in series with an AC source, as shown in Fig. 11, there are three power values to be considered. We will find a true power in *watts,* a reactive power in *vars,* and

186

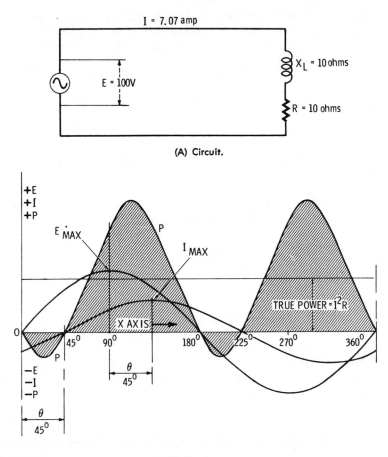

(A) Circuit.

(B) Sine-wave.

Fig. 11. Power in a circuit containing L and R in series.

an apparent power in *volt-amperes*. These power values are found as follows:

1. The impedance of the circuit is 14.14 ohms.
2. The current is 100/14.14, or 7.07 amperes.
3. The true power is I^2R, or $50 \times 10 = 500$ watts.
4. The reactive power is I^2X, or $50 \times 10 = 500$ vars.
5. The apparent power is EI, or $100 \times 7.07 = 707$ volt-amperes.

187

Only the true power does useful work in the circuit. For example, if the resistor represents a lamp filament, as in Fig. 1, the true power produces light. The reactive power merely surges back and forth in the circuit. The apparent power provides reactive power and true power, as shown in the power diagram of Fig. 12.

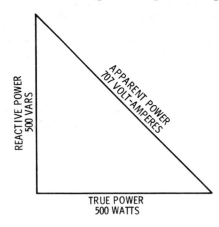

Fig. 12. A power diagram.

CAPACITIVE REACTANCE

Electricians often work with capacitive reactance, as well as with inductive reactance. Capacitance is measured in *farads* with a capacitor tester, such as illustrated in Fig. 13. A 1-farad capacitor will store 1 coulomb of charge when 1 volt is applied across the capacitor. *A coulomb* is the amount of charge produced by 1 ampere flowing for 1 second. Since the farad is a very large unit of capacitance, we generally work with *microfarads;* a microfarad is one millionth of a farad (10^{-6} farad). The symbol μ or the letter m is used to represent "micro"; for example, 50 microfarads is commonly written as $50\mu f$, $50\mu fd$, 50 mf, or 50 mfd.

We know that if an ideal inductor is connected across a DC voltage source, the inductor short-circuits the source. On the other hand, if we connect a capacitor across a DC voltage source, the capacitor is an open circuit. Although there is a momentary rush of DC current into the capacitor to charge its plates, the current flow thereafter is zero. Capacitance is the opposite of inductance in a DC circuit. We will find that they are opposites in other respects, also. We often hear electricians speak of *condensers* and *capacity*. However, it is now

188

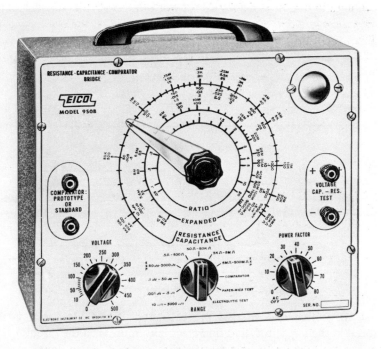

Fig. 13. Capacitance checker.

standard practice to use the terms *capacitors* and capacitance. It makes no difference which terms are used, except that the new terms are preferred in most offices and shops.

Fig. 14 shows a simple experiment of basic importance. The lamp connected in series with a capacitor and a DC voltage source does not glow. On the other hand, when an AC source is used, the lamp then glows. If we increase the capacitance, the lamp will glow brighter. By comparing Fig. 14 with Fig. 1, we will recognize that the circuit actions are opposite. When we increase the inductance in Fig. 1, the lamp glows dimmer, but when we increase the capacitance in Fig. 14, the lamp glows brighter.

It is evident that AC current does not really flow "through" a capacitor. What actually happens is the current flows into the capacitor on the

189

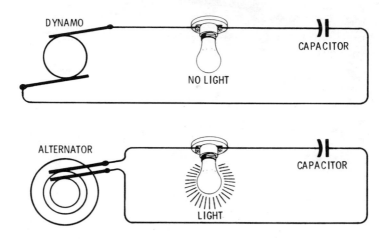

DYNAMO

CAPACITOR

NO LIGHT

ALTERNATOR

CAPACITOR

LIGHT

Fig. 14. Lamp connected in series with a capacitor.

first alternation and charges the capacitor plates. Electricity is then stored in the capacitor, and this electricity returns to the circuit on the second alternation as the capacitor discharges. We can compare a capacitor with an inductor in the sense that a capacitor does not consume power. In other words, the power in a capacitor is reactive power and is measured in vars. Therefore, the AC circuit in Fig. 14 has a reactive power value in the capacitor which is measured in *vars,* a real (true) power value in the lamp which is measured in *watts,* and an apparent power value supplied by the alternator which is measured in *volt-amperes.*

Because a large capacitance stores more electricity than a small capacitance, it also has less opposition to AC current flow. The amount of AC current flow will depend, moreover, on how many times a second the capacitor is charged and discharged.

Therefore, the AC current flow increases when the frequency is increased. In its simplest form, Ohm's law for a capacitor is written:

$$I = \frac{E}{X_c}$$

where,

I represents the number of AC amperes,
E represents the number of AC volts,
X_c represents the number of capacitive ohms.

We know that the number of capacitive ohms (reactive ohms) will depend on the number of farads of the capacitor and on the AC frequency. In turn, we write:

$$X_c = \frac{1}{2\pi f C}$$

where,

X_c represents the number of capacitive ohms,
f is the frequency in cps,
C is the capacitance in farads.

Thus, Ohm's law for a capacitor may be written in the form:

$$I = 2\pi fCE$$

For example, if a capacitor has a capacitance of 133 microfarads ($133\mu f$, or 1.33×10^{-4} farad), its reactance at 60 cps will be:

$$X_c = \frac{1}{2\pi fC} = \frac{1}{6.28 \times 60 \times 1.33 \times 10^{-4}} = 20 \text{ ohms}$$

Let us observe Fig. 15. When a sine-wave AC voltage is applied to a capacitor, the current is greatest when the voltage starts to rise from zero. This is so because there is no charge in the capacitor at this instant, and therefore there is no "back voltage" to oppose the flow of current. On the other hand, the current is zero when the voltage reaches its peak value. This is because the capacitor is then fully charged and its "back voltage" is equal and opposite to the applied peak voltage. This is just another way of saying that the current in a

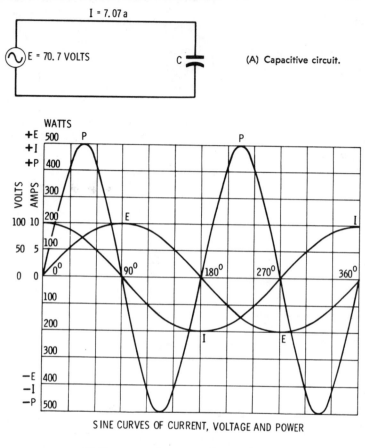

(A) Capacitive circuit.

(B) Sine curves of current, voltage and power.

Fig. 15. Voltage, current, and power in a capacitor.

capacitive circuit goes through its peak value before the applied voltage goes through its peak value.

Electricians say that the current in a capacitive circuit *leads* the applied voltage in time. This is just the opposite of an inductive circuit in which the current is delayed with respect to the applied voltage, or the current *lags* the voltage in an inductive circuit. There is a 90° phase difference between voltage and current in a capacitor. However, we have a *leading phase* in a capacitor, whereas we have a *lagging phase* in an inductor.

192

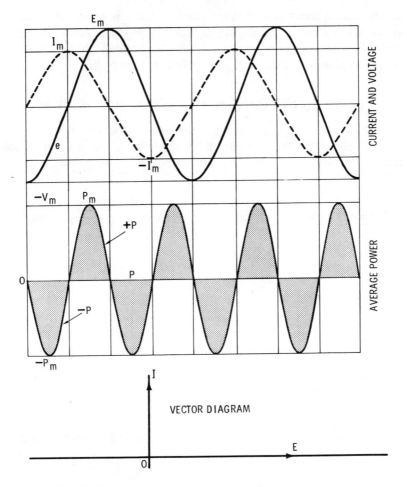

Fig. 16. The power in a capacitor is half positive and half negative.

Fig. 16 shows the power in a capacitor, which is the product of instantaneous voltages and currents. We observe that the shaded areas represent positive power and negative power on successive alternations. In other words, the average power in the capacitor is zero. All of the capacitor power is reactive power and is measured in vars:

$$\text{Vars} = E_{rms}I_{rms}$$

193

Thus, a capacitor is like an inductor in that it is a *reactor* and consumes no real power. However, capacitance is the opposite of inductance in that capacitive current leads the applied voltage by 90°, whereas inductive current lags the applied voltage by 90°.

CAPACITIVE REACTANCE AND RESISTANCE IN SERIES

Fig. 17 shows capacitance and resistance connected in series with an AC source. To find the current in this circuit, we proceed as follows:

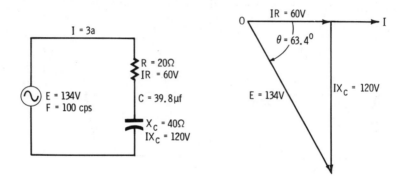

(A) Circuit.	(B) Voltage.

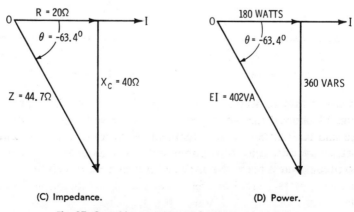

(C) Impedance.	(D) Power.

Fig. 17. Capacitive reactance and resistance in series.

1. The impedance of the circuit is $Z = \sqrt{R^2 + X^2_C} = 44.66$ ohms.
2. The current is $I = E/Z = 3$ amperes.
3. The resistive voltage drop is $E_R = IR = 60$ volts.
4. The capacitive voltage drop is $E_C = IX_C = 120$ volts.

Next, let us find the three power values in the circuit of Fig. 17:

1. The true power is $P = I^2R = 180$ watts.
2. The reactive power is $P = I^2X_C = 360$ vars.
3. The apparent power is $P = EI = 402$ volt-amperes.

Electricians use many different kinds of capacitors. For example, refrigerator electricians use capacitors such as illustrated in Fig. 18. Very

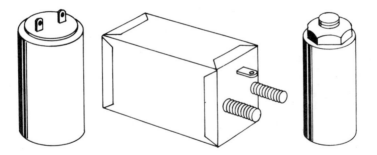

Fig. 18. Various types of motor starting capacitors.

large capacitors are used in power plants, as illustrated in Fig. 19. Capacitors which are built up from metal plates separated by sheets of insulation are called static capacitors. These capacitors operate at high voltage; for example, the capacitor in Fig. 19 is operated at a voltages which may be stated in *kilovolts;* a kilovolt is 1000 volts.

Fig. 11 showed the development of power in a circuit with inductance and resistance. The development of power in a circuit with capacitance and resistance is much the same, as seen in Fig. 20. Basically, the only difference between an RC circuit and an RL circuit is in the lead or lag of the circuit current with respect to the applied voltage in the circuit.

195

Courtesy Westinghouse Electric Corp.

Fig. 19. Large static capacitors used in factories.

INDUCTANCE, CAPACITANCE, AND RESISTANCE IN SERIES

Electricians who work with refrigerator motors and other electrical equipment are often concerned with circuits that have inductance, capacitance, and resistance connected in series with an AC voltage source, as shown in Fig. 21. As we would expect, the voltage drop across the resistor is in phase with the circuit current, while the voltage drop across the inductor is 90° out of phase with the circuit current, and the voltage drop across the capacitor is 90° out of phase with the circuit current.

Since the current is the same at any point in the circuit of Fig. 21, we take the current as a common reference in following the circuit action. In turn, the source voltage E is the vector sum of IR, IX_L, and IX_C.

196

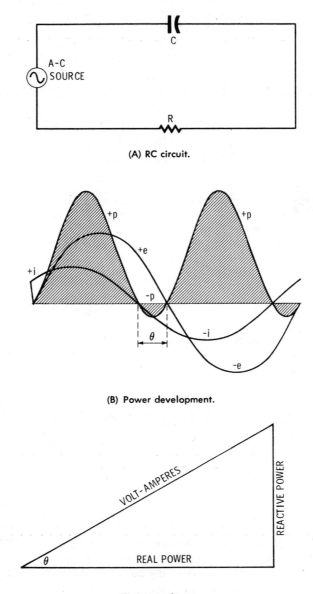

(A) RC circuit.

(B) Power development.

(C) Power diagram.

Fig. 20. Power circuit with capacitance and resistance.

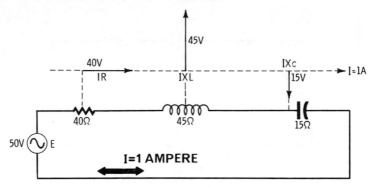

(A) Circuit.

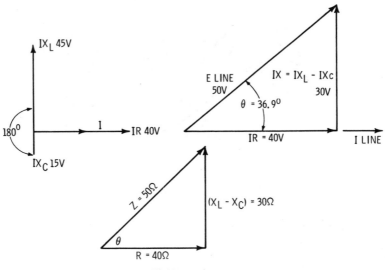

(B) Vector diagram.

Fig. 21. Resistance, inductance, and capacitance connected in series.

In Fig. 21B, we see the relations of IR, IX_L, and IX_C. Note that IX_L and IX_C are each 90° away from I, and point in opposite directions (IX_L and IX_C have a phase angle of 180°). Therefore, the total reactive voltage E_X is the difference between IX_L and IX_C. We write:

$$E_X = IX_L - IX_C = 45 - 15 = 30 \text{ volts}$$

Since X_L is greater than X_C in this example, the inductance dominates the circuit action, and the circuit current lags the applied voltage E. In other words, the phase angle in Fig. 21B is lagging. The impedance diagram for the circuit shows the impedance is 50 ohms. To find the circuit impedance by arithmetic, we write:

$$Z = \sqrt{R^2 + (X_L - X_C)^2}$$

or,

$$Z = \sqrt{40^2 + (45 - 15)^2} = \sqrt{2500} = 50 \text{ ohms}$$

We know that the circuit current in Fig. 21 is given by $I = E/Z$:

$$I = \frac{E}{Z} = \frac{50}{50} = 1 \text{ ampere}$$

In turn, the four power values in the circuit of Fig. 21 are as follows:

$$P_R = I^2 R = 40 \text{ watts}$$

$$P_L = I^2 X_L = 45 \text{ vars}$$

$$P_C = I^2 X_C = 15 \text{ vars}$$

$$P_{apparent} = EI = 50 \text{ volt-amperes}$$

Note that the difference between P_L and P_C is 30 vars. We know that apparent power is equal to:

$$P_{app} = \sqrt{\text{Real power}^2 + \text{Reactive power}^2}$$

or,

$$P_{app} = \sqrt{40^2 + 30^2} = 50 \text{ volt-amperes}$$

In large factories and power plants, electricians are particularly concerned with adjusting the capacitance in circuits such as shown in Fig.

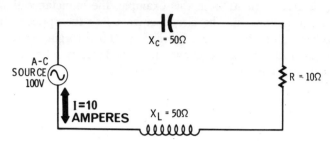

(A) X_C and X_L are equal in this circuit.

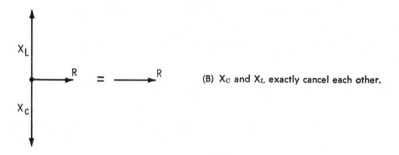

(B) X_C and X_L exactly cancel each other.

Fig. 22. Conductive and reactance circuit.

21 so that the inductive and capacitive reactances are equal. Fig. 22 shows a circuit in which $X_L = X_C$. We recognize that X_L exactly cancels X_C. Therefore, *the circuit action is the same as though there were only resistance present in the circuit.* Since there is effectively no inductance or capacitance in this circuit, the phase angle between the source voltage and the circuit current is zero. Electricians say that this kind of a circuit has been *corrected for zero power factor.*

It follows that the current in the circuit of Fig. 22 is given by:

$$I = \frac{E}{R} = \frac{100}{10} = 10 \text{ amperes}$$

When the inductive reactance is equal to the capacitive reactance in a circuit, we see that the number of volt-amperes is the same as the resistive load power. This is a desirable condition in a power-distribution system because there is no reactive current surging in the line to increase the line losses. Therefore, the efficiency of a power-distribu-

200

tion system is greatest when the inductive reactance that may be present is exactly cancelled by inserting an equal amount of capacitive reactance. This is the purpose of the large capacitor illustrated in Fig. 19.

Some electricians call the circuit in Fig. 22 a *series-resonant circuit*. This simply means that the phase angle between the source voltage and circuit current is zero. It is important to note that there is a *voltage rise* across the capacitor and across the inductor in a series-resonant circuit. For example, in Fig. 22, the voltage drop across the capacitor is equal to IX_C, or 500 volts. Similarly, the voltage drop across the inductor is equal to IX_L, or 500 volts. Although the source voltage is 100 volts in Fig. 22, the voltage across L rises to 500 volts and the voltage across C rises to 500 volts. In turn, we must use inductors and capacitors that are rated for 500-volt operation. This voltage rise across an inductor and across a capacitor in a resonant LCR circuit

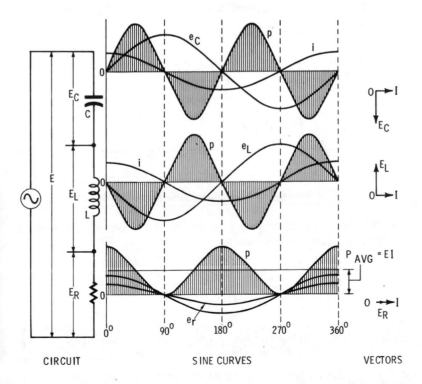

CIRCUIT   SINE CURVES   VECTORS

Fig. 23. Instantaneous voltage, current, and power relations in a series-resonant circuit.

201

is sometimes called the *voltage magnification* of the circuit. The voltage magnification can be very great when the load resistance is small. For example, if we used a 1-ohm load resistor in Fig. 22, the voltage across L would be 5000 volts, and across C would be 5000 volts. Fig. 23 shows the instantaneous voltage, current, and power relations in a series-resonant circuit.

INDUCTANCE AND RESISTANCE IN PARALLEL

Electricians often work with circuits that consist of inductance and resistance connected in parallel. For example, an AC line may be connected to a group of lamps and to a motor such as an induction motor. The lamps provide a resistive load, and the motor provides a load which has an inductive component. Let us consider the impedance of an RL parallel circuit, such as shown in Fig. 24.

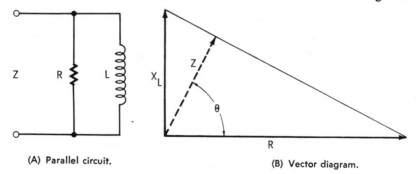

(A) Parallel circuit.   (B) Vector diagram.

Fig. 24. Impedance parallel circuit.

The inductive reactance has a value of $2\pi fL$ ohms. To find the circuit impedance, we may write:

$$Z = \frac{RX_L}{\sqrt{R^2 + X_L^2}}$$

However, to simplify the calculation of Z, it is advisable to make use of the triangle relations shown in Fig. 24B. We lay off the line R with a length proportional to the resistance, and lay off the line X_L with a length proportional to $2\pi fL$. Then, we draw the line Z perpendicular to the hypotenuse of the triangle. The length of Z then gives the number of ohms of impedance. Note also that the angle θ is the power-factor angle of the RL parallel circuit.

When working with power-factor correction, as will be explained, electricians often find it useful to change a parallel RL circuit into an equivalent series RL circuit, as shown in Fig. 25. In other words, when an inductance L_P is connected in parallel with a resistance R_p (Fig. 25A), we can make up an equivalent circuit by connecting

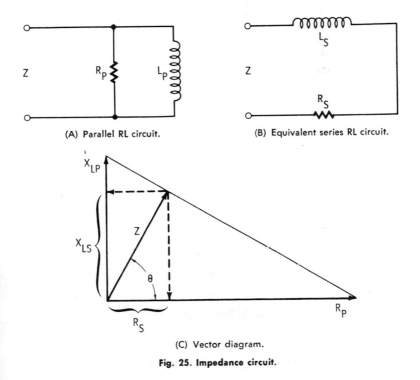

(A) Parallel RL circuit.

(B) Equivalent series RL circuit.

(C) Vector diagram.

Fig. 25. Impedance circuit.

an inductance L_s in series with a resistance R_s (Fig. 25B). The relations of inductance and resistance values in the series and parallel circuits are easily found by drawing the triangle relations shown in Fig. 25C. Note that smaller values of series inductance and resistance have the same circuit action as larger values of parallel inductance and resistance.

CAPACITANCE AND RESISTANCE IN PARALLEL

Electricians may also work with circuits that consist of capacitance and resistance connected in parallel. For example, an AC line may be connected to a group of lamps and to a motor such as a

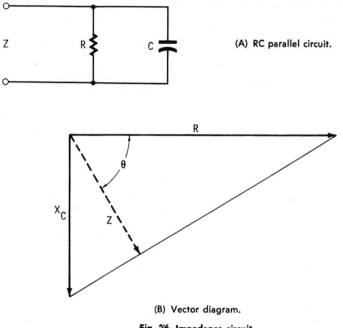

(A) RC parallel circuit.

(B) Vector diagram.

Fig. 26. Impedance circuit.

synchronous motor. The lamps provide a resistive load, and the motor provides a load which has a capacitive component. Let us consider the impedance of an RC parallel circuit, such as shown in Fig. 26. The capacitive reactance has a value of $1/2\pi fC)$ ohms. To find the circuit impedance, we may write:

$$Z = \frac{RX_C}{\sqrt{R^2 + X_C^2}}$$

However, to simplify the calculation of Z, it is advisable to make use of the triangle relations shown in Fig. 26B. We lay off the line R with a length proportional to the resistance, and lay off the line X_C with a length proportional to $1/(2\pi fC)$. Then, we draw the line Z perpendicular to the hypotenuse of the triangle. The length of Z then gives the number of ohms of impedance. Note also that the angle θ is the power-factor angle of the RC parallel circuit.

When working with power-factor correction, electricians often find it useful to change a parallel RC circuit into an equivalent series RC circuit, as shown in Fig. 27. In other words, when a capacitance C_p is connected in parallel with a resistance R_p, we can make up an equivalent circuit by connecting a capacitance C_s in series with

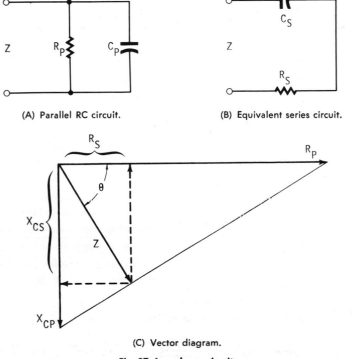

(A) Parallel RC circuit.   (B) Equivalent series circuit.

(C) Vector diagram.

Fig. 27. Impedance circuit.

a resistance R_s. The relations of capacitance and resistance values in the series and parallel circuits are easily found by drawing the triangle relations shown in Fig. 27C.

205

INDUCTANCE, CAPACITANCE, AND RESISTANCE IN PARALLEL

Electricians in factories or shops are sometimes concerned with circuits that contain resistance, capacitance, and inductance in parallel. For example, an AC line may be connected to a group of lamps, to a synchronous motor, and to an induction motor. In this case, we are concerned with a basic RCL parallel circuit as shown in Fig. 28. The R, X_L, and X_C vectors are drawn as shown in

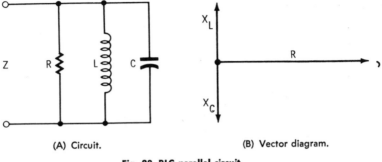

(A) Circuit.   (B) Vector diagram.

Fig. 28. RLC parallel circuit.

Fig. 28B. Note that X_L and X_C extend in opposite directions. Therefore, X_L and X_C will tend to cancel each other, and if $X_L = X_C$, the reactances cancel completely, leaving resistance only.

When the values of X_L and X_C are equal, electricians state that the circuit in Fig. 28A is a *parallel-resonant circuit.* This means that L and C are effectively not present, inasmuch as they cancel each other. Therefore, current is drawn by R only. When a circuit is parallel-resonant, it is also said to be *power-factor corrected,* since the power-factor angle is obviously zero in the power-factor corrected circuit, particularly when the power demand is heavy. This is because a power-factor corrected circuit has no surging line currents, and therefore operates at maximum efficiency. (The I^2R line loss due to surging line currents has been eliminated.)

From a practical point of view, induction motors are in wider use than synchronous motors. If we add a suitable number of synchronous motors to a group of induction motors, we can operate the power circuit at unity power factor. This is the most common

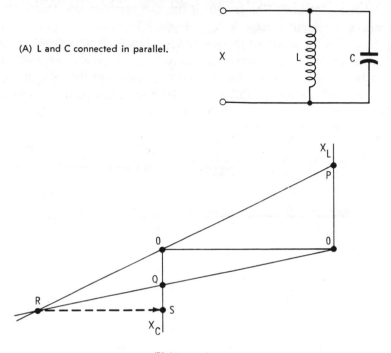

(A) L and C connected in parallel.

(B) Vector diagram.

Fig. 29. Total reactance.

method of obtaining power-factor correction. However, we can also connect a capacitor of suitable value across an induction motor, and the motor will operate at unity power factor. Such a capacitor is often called a static capacitor, as noted previously.

When power-factor correction is partial (incomplete), there is a reactance left over in the circuit. For example, let us consider the LC parallel circuit shown in Fig. 29A. In this example, X_L is greater than X_C, and there is a capacitive reactance left over in the circuit. The input reactance X is a capacitive reactance. To find out how much capacitive reactance is left over, we use the triangle relations shown in Fig. 29B. Point P on the X_L axis represents the amount of inductive reactance; point Q on the X_C axis represents the amount of capacitive reactance. The lines PO and OQ intersect at point R. We draw a line RS perpendicular to the X_C axis, thereby

207

determining point S. The length of the line segment OS denotes the value of the input reactance X in Fig. 29A.

Note that the length of the line O,O in Fig. 29B must be 10 units. That is, we lay off this horizontal line equal to 10 reactance units. It is only necessary that we make OP proportional to the inductive reactance of L, and make OQ proportional to the capacitive reactance of C.

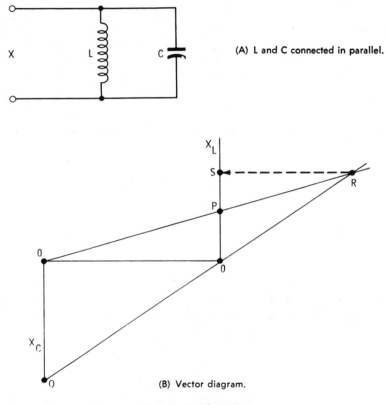

(A) L and C connected in parallel.

(B) Vector diagram.

Fig. 30. Total reactance.

Note that if X_C should happen to be greater than X_L, the diagram would be drawn as shown in Fig. 30. In this example, there is inductive reactance left over, and the amount of inductive reactance that is left over is given by the length of the line segment OS.

With these principles in mind, it is easy to find the amount of capacitance (or inductance) that will be required to correct the power factor in any practical circuit. For example, suppose that an induction motor consists of an effective inductance in series with an effective resistance, as depicted in Fig. 25B. In turn, the triangle diagram in Fig. 25C shows the amount of effective parallel inductance. To correct the power factor, we must connect a suitable amount of capacitance in parallel with the inductance. In other words, the reactance of this capacitance must be equal to the reactance of the effective parallel inductance that we have determinded.

SUMMARY

Inductance is the property which opposes any change in the existing current. Inductance opposes any change in current flow. Inductance is measured in henrys. Inductance is the result of magnetic lines of force cutting the coil turns when magnetic field strength changes. Therefore, the inductance of an air-core coil is greatly increased if we insert an iron core.

The total opposition to resistance and inductive reactance is called impedance. Impedance is measured in ohms. Another type of reactance is called capacitive reactance, which is also measured in ohms. A capacitor is like an inductor in that it is a reactor and consumes no real power. Capacitance is the opposite of inductance in that capacitive current leads the applied voltage by 90° whereas inductive current lags the applied voltage by 90°.

A series-resonant circuit is one in which an inductor and capacitor are connected in series and have values such that the inductive reactance of the inductor will be equal to the capacitive reactance of the capacitor at a particular frequency.

TEST QUESTIONS

1. How can the brightness of a lamp be controlled with a variable inductance?
2. Explain the different symbols that may be used for coils and resistors by electricians.
3. How is a noninductive coil wound? Where would it be used?

4. State Ohm's law for a purely inductive circuit.
5. In what way does true power differ from reactive power?
6. Discuss how to find the reactance of a coil.
7. How do we define impedance?
8. Explain how to find the impedance of a circuit that has resistance and inductance.
9. What is meant by a power diagram?
10. Discuss how to find the reactance of a capacitor.
11. Define a henry. A farad.
12. What is the definition of phase angle?
13. Explain the difference between lagging and leading phase angles.
14. How do we draw an impedance diagram?
15. What is an instrument called which measures real power? Reactive power?
16. How would we measure the apparent power in a reactive circuit?
17. Why does capacitive reactance tend to cancel inductive reactance in a series circuit?
18. If the inductive reactance is exactly equal to the capacitive reactance in an LCR series circuit, how do we describe the circuit?
19. What is the value of the power-factor angle in a resonant circuit?
20. Explain what electricians mean by power-factor correction.
21. Why is power-factor correction important in power-distribution systems?
22. Discuss an example of voltage rise across an inductor and across a capacitor in a series-resonant circuit.
23. Why is the voltage magnification greater in a series-resonant circuit when the load resistance is reduced in value?
24. Why is voltage magnification of concern to the electrician?
25. What is an instrument called which measures capacitance values?

CHAPTER 8

Electric Lighting

Light is a form of energy. It has wave-like properties and is called a visible form of radiant energy. Light and heat are basically very similar; however, heat has a longer wavelength than light, and is invisible. Only those radiant wavelengths between the approximate limits of 0.00038 to 0.00076 millimeters are visible to the human eye. Note that a meter contains 39.37 inches, and that a millimeter is 0.001 of a meter. For convenience, the wavelengths of light are usually expressed in *angstroms*. One angstrom is equal to 1/10,000,000 of a millimeter. In other words, the range of visible wavelengths is from about 3800 to 7600 angstroms. Table 1 shows the visible wavelengths which are a very small "slice" in the extensive range of radiant energy.

Table 1. Classifications of Radiant Energy

Frequency, cycles/sec										
10^2	10^4	10^6	10^8	10^{10}	10^{12}	10^{14}	10^{16}	10^{18}	10^{20}	10^{22}
10^6	10^4	10^2	1	10^{-2}	10^{-4}	10^{-6}	10^{-8}	10^{-10}	10^{-12}	10^{-14}
Wavelength, meters										

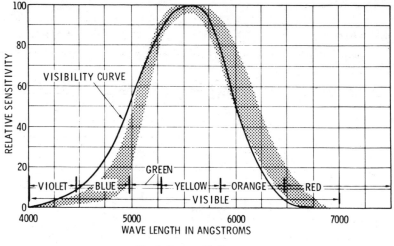

Fig. 1. Eye sensitivity curve.

Within the foregoing visible range, the colors corresponding to various wavelengths are as shown in Fig. 1. The eye has greatest sensitivity for wavelengths of about 5500 angstroms; that is, yellow can be seen under such poor conditions of illumination when blue or red cannot be seen. Under dim illumination, the sensitivity curve tends to shift as shown by the shaded region in Fig. 1. Therefore, violet disappears first and red remains visible. Yellow disappears last as the illumination becomes very dim. As each color disappears, it becomes a gray shade, and finally black.

SOURCES OF LIGHT

Light sources are broadly divided into natural sources such as the sun, moon, stars, Aurora Borealis, phosphorescent insects and fishes, and open fires. The other broad division is into sources of artificial light, such as oil and gas lamps, incandescent lamps, arc lamps, various types of gaseous glow lamps, and fluorescent lamps. Fig. 2 provides a brief survey of both natural and artificial light sources.

MEASUREMENT OF AMOUNT OF ILLUMINATION

The original method of comparing amounts of illumination was to use the light from a standard tallow candle held 12 inches from the sur-

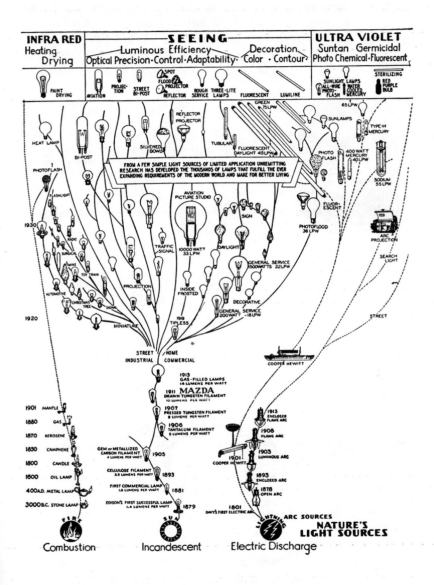

Courtesy General Electric Co.

Fig. 2. Pictorial summary of light.

213

face to be illuminated. This unit of illumination is called the *foot-candle*. Electricians often use a light meter to measure foot-candles, as illustrated in Fig. 3. We are also concerned in many practical situations with the total light output that a lamp radiates in all directions.

Fig. 3. Lightmeter equipped with a light sensitive cell.

Courtesy Sargent-Welch Scientific Co.

Let us consider a point source of light placed inside a sphere that has a radius of 12 inches as shown in Fig. 4. If the entire inner surface of the sphere is illuminated by an amount of 1 foot-candle, the total light output from the source is *1 lumen*.

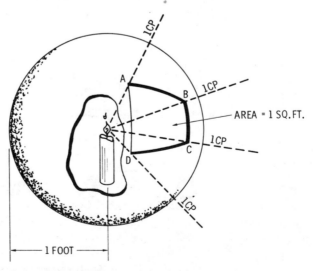

Fig. 4. Relation of candles to lumens.

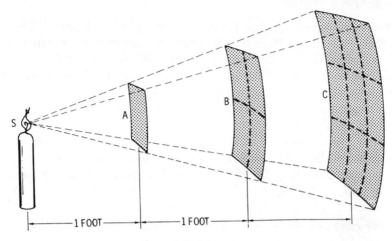

Fig. 5. Illustrating the inverse square law.

We will find that the intensity of illumination falls off as the square of the distance from the source. With reference to Fig. 5, note that the illumination at *A* is 1 foot-candle. Next, if we remove surface *A* and illuminate surface *B,* the same amount of light now illuminates 4 times as much area. Therefore, the illumination at *B* is ¼ foot-candle. Then, if we remove surface *B* and illuminate surface *C,* the same amount of light now illuminates 9 times as much area as at *A*. Therefore, the illumination at *C* is 1/9 foot-candle.

To get a basic idea of the amount of illumination provided by an incandescent lamp, it is helpful to note that a 60-watt lamp with a tungsten filament and an inside-frosted bulb has a brightness of about 60 foot-candles. By way of comparison, the noon-day sun has a brightness of about 1,000,000 foot-candles. The moon has a brightness of approximately 5 foot-candles. In other words, these are comparative brightnesses of the light sources. Very bright light sources such as the sun or an arc lamp will damage the retina of the eye if viewed directly.

TUNGSTEN FILAMENTS

Since the light output from a filament depends on its temperature, only metals with high melting points are suitable for use as lamp filaments. Tungsten melts at approximately 6120°F. A typical tungsten lamp is operated at a filament temperature of 4800°F. Since tungsten gradually evaporates at this temperature, the average life of the lamp is about 1000 hours. Note that if a tungsten filament is operated 5% below rated voltage, the lamp life will be doubled, but the light output will be 17% below normal. The efficiency of the lamp (light output divided by power input) will be reduced 10%.

Tungsten has the best efficiency of any filament material, and 20 lumens per watt is typical for large lamps. A 100-watt tungsten lamp will have an efficiency of about 15 lumens per watt. A 250-watt photoflood lamp has an efficiency of approximately 35 lumens per watt, but its average life is only 3 hours. A 6-watt lamp, such as used in electric signs, has an average life of 1500 hours, but its efficiency is less than 7 lumens per watt. Filaments in high-vacuum lamps are operated at about 4000°F; gas-filled lamp filaments operate 800° higher, and photoflood lamp filaments operate a little above 5600°F.

A gas-filled lamp tends to reduce evaporation of the tungsten filament and can therefore be operated at a higher temperature than a vacuum lamp. Most gas-filled lamps employ a mixture of nitrogen and argon. Krypton is the best gas for this purpose but is so expensive that it is used only in special-purpose lamps, such as miner's lamps. The gas pressure in a gas-filled lamp is about 20% below atmospheric pressure. Most filaments are *coiled* in the form of a long helical spring which reduces the tendency of the gas to cool the filament. A helical filament is sometimes coiled into another helix of greater diameter to further reduce the cooling action of the gas.

Helical filaments also have a slower rate of tungsten evaporation. This evaporation eventually causes *bulb blackening* because the tungsten vapor condenses as a black film on the inner surface of the bulb. In a gas-filled lamp, the hot gas carries the tungsten vapor upward. Therefore, a black spot forms at the top of the bulb instead of spreading over the entire inner surface, as in a high-vacuum bulb. Chemicals, called "getters," are often placed inside the bulb to capture tungsten vapor and thereby reduce the rate of blackening. A piece of wire mesh called a *collector grid* may also be attached to each lead-in wire to attract the particles of tungsten vapor.

STARTING SURGE OF CURRENT

We know that a metal increases in resistance as its temperature increases. Fig. 6 shows a current-vs-voltage curve for a tungsten fila-

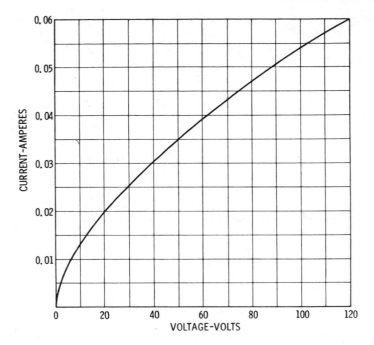

Fig. 6. Current versus voltage curve for a tungsten filament rated for 120 volt operation.

ment. The current starts to rise rapidly at first, and then rises more slowly as the filament becomes hotter. Fig. 7 shows how the resistance of a tungsten filament increases as it heats up. The "cold" resistance of an ordinary lamp is about 1/15 of its "hot" resistance. The "cold" resistance of a photoflood lamp is approximately 1/20 of its "hot" resistance. Since it takes a certain amount of time for a filament to come up to operating temperature, a surge of current flows when the switch is turned on. This starting surge of current is 15 times as great as the normal operating current for an ordinary lamp, and is 20 times as great for a photoflood lamp.

The practical electrician must use a switch that will withstand starting surges without damage. For example, in a shop or factory, a line of lamps might draw 50 amperes in normal operation. However, the starting-surge current will be 50×15, or 750 amperes. Unless the switch contacts can withstand 750 amperes without damage, the life of the switch will be comparatively short. As another example, if a line

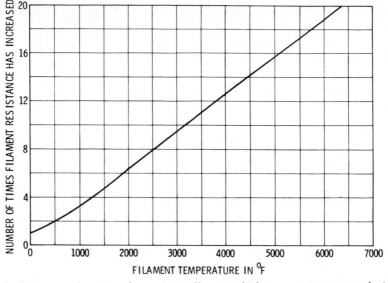

Fig. 7. Increase of resistance by a tungsten filament at high temperatures, compared with its resistance at 70°F.

of photoflood lamps normally operates at 50 amperes, the switch contacts must withstand 50×20, or 1000 amperes without damage. Otherwise, the switch is likely to fail after a few operations.

AGING CHARACTERISTICS

As a tungsten lamp ages, its light output decreases for two reasons. Evaporation of the filament tends to cause the bulb to blacken. Also, evaporation makes the filament slowly decrease in diameter, which means that the *filament resistance increases*. Therefore, an old filament draws less current and operates at a lower temperature, which reduces its light output. In turn, the efficiency of the lamp (in lumens per watt) also decreases with age. The current drawn and the power consumed by the filament decrease at the same rate as the lamp ages. However, the efficiency decreases about four times as fast, and the light output decreases approximately five times as fast.

Electricians use a "rule of thumb" which states that when a lamp ages to a point that its light output has decreased 30%, it should be washed to remove any dust, dirt, or grease that may have accumulated

218

on its outer surface. As a lamp ages, it may also reach a so-called "smashing point," even if the filament does not burn out. This is defined as the point at which the light output of the lamp has fallen sufficiently that the cost of electricity per 1,000,000 lumen-hours is greater than the average cost of light produced up to that time, plus the initial cost of the lamp and electricity. The "smashing point" is of practical importance only in large lighting systems where operating costs are comparatively high.

LAMP BASES AND BULBS

Various types of lamp bases are used in practice. The *screw-type* base shown in Fig. 8 is the most common. The smallest size is called the *miniature* base, followed by the *candelabra, intermediate, medium,*

Fig. 8. Various types of screw base used in incandescent lamps.

and the *mogul* bases. The medium base is used in ordinary home-lighting systems. Other types of bases include the *bayonet, prefocus,* and *bipost* bases, as shown in Fig. 9. Bayonet bases are made in a number of different sizes, and have two projecting side pins which seat the base in its socket. Automotive electricians use the *bayonet candelabra* size of base in automobile lamps.

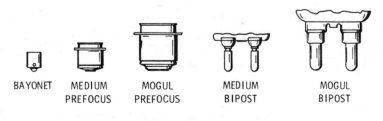

BAYONET   MEDIUM   MOGUL   MEDIUM   MOGUL
           PREFOCUS   PREFOCUS   BIPOST   BIPOST

Fig. 9. Bayonet, prefocus, and bipost bases.

219

Table 2. Miniature Lamp Data

Lamp No.	Volts	Amps	Bead Color	Base	Bulb Type	Fig. No.
PR2	2.4	0.50	Blue	Flange	B-3½	A
PR3	3.6	0.50	Green	Flange	B-3½	A
PR4	2.3	0.27	Yellow	Flange	B-3½	A
PR6	2.5	0.30	Brown	Flange	B-3½	A
PR12	5.95	0.50	White	Flange	B-3½	A
12	6.3	0.15	— — —	2-Pin	G-3½	H
13	3.8	0.30	Green	Screw	G-3½	B
14	2.5	0.30	Blue	Screw	G-3½	B
40	6.3	0.15	Brown	Screw	T-3¼	C
41	2.5	0.50	White	Screw	T-3¼	C
42	3.2	0.35*	Green	Screw	T-3¼	C
43	2.5	0.50	White	Bayonet	T-3¼	D
44	6.3	0.25	Blue	Bayonet	T-3¼	D
45	3.2	0.35†	Green†	Bayonet	T-3¼	D
46	6.3	0.25	Blue	Screw	T-3¼‡	C
47	6.3	0.15	Brown	Bayonet	T-3¼	D
48	2.0	0.06	Pink	Screw	T-3¼	C
49	2.0	0.06	Pink	Bayonet	T-3¼	D
50	6.3	0.20	White	Screw	G-3½	B
51	6.3	0.20	White	Bayonet	G-3½	E
55	6.3	0.40	White	Bayonet	G-4½	F
57	14.0	0.24	White	Bayonet	G-4½	F
112	1.1	0.22	Pink	Screw	TL-3	G
222	2.2	0.25	White	Screw	TL-3	G
233	2.3	0.27	Purple	Screw	G-3½	B
291	2.9	0.17	White	Screw	T-3¼	C
292	2.9	0.17	White	Screw	T-3¼	C
1490	3.2	0.16	White	Bayonet	T-3¼	D
1891	14.0	0.23	Pink	Bayonet	T-3¼	D
1892	14.0	0.12	White	Screw	T-3¼	C

* Some brands are .50 amp.
† Some brands are .50 amp and white bead.
‡ Frosted.

The prefocus base has two side-flange projections and is made in medium and mogul sizes. It is actually a double base which is assembled at the factory so that when it is seated in its socket, the filament will have an exact position with respect to the optical system with which it is used. This feature is important for projector lamps. A bipost base has two metal pins sealed into the lower portion of the bulb. It also provides an exact position for the filament in an optical system.

Table 2 lists the most common miniature lamps and their characteristics. Outline drawings for each lamp are shown in Fig. 10. Miniature lamps are widely used in electrical equipment such as instrument

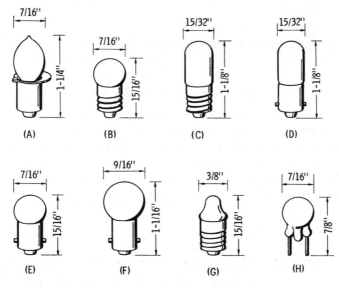

Fig. 10. Outline drawings for miniature lamps.

panels, flashlights, off-on indicators, etc. It is important for the electrician to observe the bead color-code when replacing a miniature lamp. In other words, the color code indicates the voltage and current rating of the lamp.

Bulbs of larger lamps are made in various types and are described by letters, as shown in Fig. 11. The bulbs may be clear, colored, inside-frosted, or coated with diffusing or reflecting materials. In addition to its letter description, a bulb is also identified by a number which indicates its diameter in eighths of an inch. For example, a T-24 bulb tells

221

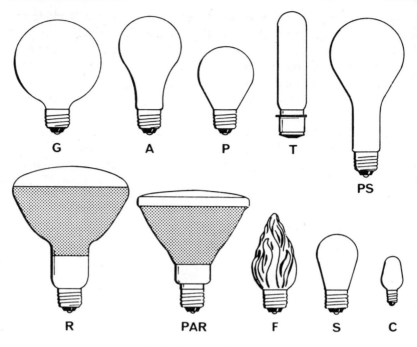

Fig. 11. Bulb-type designations.

the electrician that the lamp has a tubular shape and is 3 inches in diameter.

Lamps are specified in overall length from the top of the bulb to the bottom of the base. The *light-center length* is measured from the center of the filament to the bottom of a screw base; or to the top of the base pins or flanges of a bayonet or prefocused base; or to the shoulder of the post on a mogul bipost base; or to the bottom of the bulb (base end) of a medium bipost base.

Inside frosting is often chosen in selecting a bulb because it gives a moderate *diffusion* of the light, thereby reducing glare and shadow intensity. Since the outer surface of the bulb is smooth, it can be easily cleaned. Inside frosting results in a slight loss of light—usually between 1% and 2%. Some bulbs are made of white glass (milky appearance), or are white-coated; these bulbs provide greater diffusion of the light, but also lose from 15% to 20% of the light output from the filament. Electricians sometimes prefer high-vacuum lamps instead

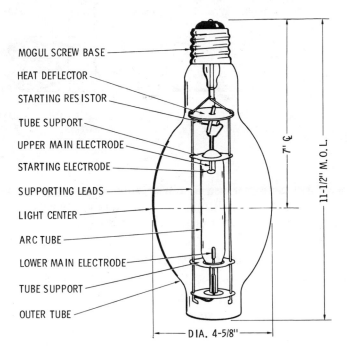

MOGUL SCREW BASE
HEAT DEFLECTOR
STARTING RESISTOR
TUBE SUPPORT
UPPER MAIN ELECTRODE
STARTING ELECTRODE
SUPPORTING LEADS
LIGHT CENTER
ARC TUBE
LOWER MAIN ELECTRODE
TUBE SUPPORT
OUTER TUBE

7" C
11-1/2" M.O.L.
DIA. 4-5/8"

(A) Mercury-vapor.

(B) Neon glow bulb.

Fig. 12. Typical bases used on vapor-discharge lamps.

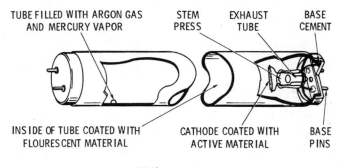

TUBE FILLED WITH ARGON GAS AND MERCURY VAPOR — STEM PRESS — EXHAUST TUBE — BASE CEMENT

INSIDE OF TUBE COATED WITH FLOURESCENT MATERIAL — CATHODE COATED WITH ACTIVE MATERIAL — BASE PINS

(C) Fluorescent.

of gas-filled lamps for use in refrigeration rooms, because gas-filled lamps produce more heat. If meat is hung too close to a gas-filled lamp, it may be warmed enough to cause early spoiling.

VAPOR-DISCHARGE LAMPS

Vapor discharge lamps produce light by means of the flow of electricity through gases. Among the more important types are the *mercury, sodium, neon, neon glow, fluorescent,* and *ultraviolet germicidal* lamps. Fig. 12 shows some familiar vapor-discharge lamps and bases. Vapor-discharge lamps are generally classified as *hot-filament, low-voltage* types, *high-voltage, cold-cathode* types, *and mercury-pool arc, low-voltage* types. The mercury lamp is a hot-cathode, low-voltage type that operates in the range from 100 to 200 volts. It produces light by conduction of electricity through mercury vapor and has a blue-violet light output. This type is used in photographic work and for high-intensity illumination for use in industrial plants.

The sodium type of lamp, shown in Fig. 13, is a hot-cathode, low-voltage type that operates in the range from 100 to 200 volts. It produces light by conduction of electricity through sodium vapor and has a yellow light output. Sodium lamps are used for lighting bridges and highways. The neon type of lamp has a cold cathode and operates in the voltage range from 3000 to 15,000 volts It is formed from glass tubes that contain neon gas, and may be quite long. Neon gas produces orange-red light; this type of lamp is used in commercial signs and for display lighting.

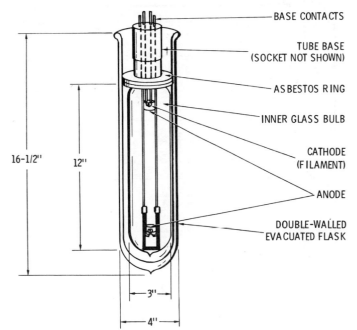

BASE CONTACTS

TUBE BASE
(SOCKET NOT SHOWN)

ASBESTOS RING

INNER GLASS BULB

CATHODE
(FILAMENT)

ANODE

DOUBLE-WALLED
EVACUATED FLASK

16-1/2"   12"

3"

4"

Fig. 13. Construction of a sodium lamp.

A neon glow lamp has a cold cathode which is treated with special materials that provide conduction of electricity in the voltage range from 60 to 200 volts. These lamps have a power consumption from 0.5 to 3 watts and produce an orange-red glow. They are used chiefly as indicators, in electrical test equipment, and as night-lights or marker lights. Germicidal lamps use hot cathodes with mercury vapor in special glass envelopes that can pass ultraviolet light. They are used to sterilize the air in their vicinity.

Ordinary fluorescent lamps used in factories, shops, and homes are made with glass tubes which are coated on their inner surface with chemical powders that fluoresce when struck by ultraviolet light. When this type of lamp is filled with neon gas, it is called a hot-cathode, low-voltage neon lamp. Commercial signs and displays frequently use the high-voltage mercury, long-tube type of lamp which operates in the range from 3000 to 15,000 volts. They are similar to high-voltage neon lamps except that mercury vapor is added to pro-

225

Table 3. Gas-Filled Lamps

Number	Hours of Average Useful Life*	Type Gas	Max. Length In inches	Base	Amps	Volts	Watts†
AR-1	3,000	Argon	3 1/2	Medium Screw	0.018	110-125	2
AR-3	1,000	Argon	1 5/8	Cand. Screw	0.0035	110-125	1/4
AR-4	1,000	Argon	1 1/2	Double-Contact Bayonet	0.0035	110-125	1/4
NE-2	Over 25,000	Neon	1 1/16‡	Unbased	0.003	110-125	1/25
NE-2A	Over 25,000	Neon	27/32‡	Unbased	0.003	110-125	1/25
NE-17	5,000	Neon	1 1/2	Double-Contact Bayonet§	0.002	110-125	1/4
NE-30	10,000	Neon	2 1/4	Medium Screw§	0.012	110-125	1
NE-32	10,000	Neon	2 1/16	Double-Contact Bayonet§	0.012	110-125	1
NE-34	8,000	Neon	3 1/2	Medium Screw	0.018	110-125	2
NE-40	8,000	Neon	3 1/2	Medium Screw§	0.030	110-125	3
NE-45	Over 7,500	Neon	1 5/8	Cand. Screw	0.002	110-125	1/4
NE-48	Over 7,500	Neon	1 1/2	Double-Contact Bayonet	0.002	110-125	1/4
NE-51	Over 15,000	Neon	1 3/16	Miniature Bayonet	0.0003	110-125	1/25
NE-56	10,000	Neon	2 1/4	Medium Screw§	0.005	220-250	1
NE-57	5,000	Neon	1 5/8	Cand. Screw§	0.002	110-125	1/4
NE-58	Over 7,500	Neon	1 5/8	Cand. Screw	0.002	220-250	1/2

* Life on DC is approximately 60% of AC values.
† For 110-125V operation.
‡ The dimension is for glass only.
§ On DC circuits the base should be negative.

Table 4. External Resistances Needed for Gas-Filled Lamps

Type	110-125V	220-300V	300-375V	375-450V	450-600V
AR-1	Included in Base	10,000	18,000	24,000	30,000
AR-3	Included in Base	68,000	91,000	150,000	160,000
AR-4	15,000	82,000	100,000	160,000	180,000
NE-2	200,000	750,000	1,000,000	1,200,000	1,600,000
NE-2A	200,000	750,000	1,000,000	1,200,000	1,600,000
NE-17	30,000	110,000	150,000	180,000	240,000
NE-30	Included in Base	10,000	20,000	24,000	36,000
NE-32	7,500	18,000	27,000	33,000	43,000
NE-34	Included in Base	9,100	13,000	16,000	22,000
NE-40	Included in Base	6,200	8,200	11,000	16,000
NF-45	Included in Base	82,000	120,000	150,000	200,000
NE-48	30,000	110,000	150,000	180,000	240,000
NE-51	200,000	750,000	1,000,000	1,200,000	1,600,000
NE-56	Included in Base	—	—	—	—
NE-57	Included in Base	82,000	120,000	150,000	200,000
NE-58	Included in Base	—	—	—	—

duce blue light. The neon gas in these lamps is used only to provide easy starting. If this type of lamp is coated on the inner surface with fluorescent powders, it is called a high-voltage fluorescent lamp.

All vapor-discharge lamps operate on the basis of arc discharge and will be destroyed by excessive current flow unless connected in series with a resistor or reactor called a *ballast*. Table 3 lists various types of neon and argon lamps. Table 4 lists the values of resistance that are connected in series with each type. Vapor-lamp ballasts may consist of a series resistor, or a reactor such as an inductor, a capacitor, or a transformer that has comparatively high leakage reactance. A ballast resistor drops from ⅓ to ⅔ of the line voltage and therefore wastes considerable power.

Power loss can be minimized by using an inductor as a ballast. Although there is some power loss in the winding resistance, most of the voltage drop occurs across the inductive reactance and consumes no true power. The power-factor angle is about 45°, which means that

227

reactive power surges back-and-forth in the line and increases the line losses. If a capacitor is used as a ballast, the power-factor angle is also about 45°, but the current leads instead of lagging as with an inductor. Electricians sometimes use a capacitor ballast with one lamp and an inductor ballast with another lamp so that the total power-factor angle is practically zero.

Starters are used with fluorescent lamps, as shown in Fig. 14. The starter consists of a neon-bulb thermostat. When the 117-volt AC input is applied, the neon gas conducts electricity and glows. This

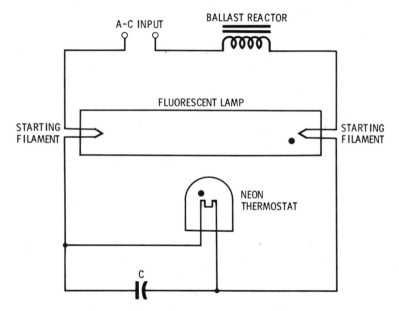

Fig. 14. The neon thermostat operates as a starter.

produces heat which causes the thermostat contacts to close. Then, the starting filaments in the fluorescent lamp heat up. The neon glow stops because the thermostat contacts are closed in the neon bulb. This cools the thermostat, and its contacts open. Opening of the filament circuit causes a high-voltage inductive kickback from the ballast reactor which starts the arc discharge in the fluorescent lamp. The voltage drop across the lamp (arc drop) is too small to fire the neon bulb. Therefore, the thermostat contacts remain open. Capacitor C in Fig. 14 minimizes sparking at the thermostat contacts.

228

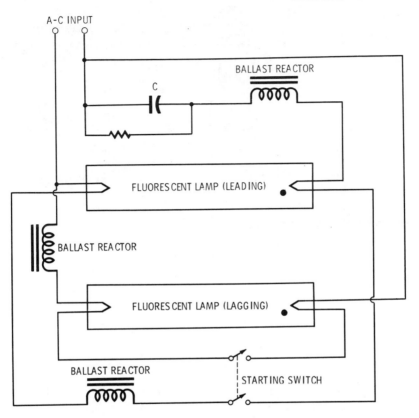

Fig. 15. A capacitor in the upper ballast circuit reduces strobe flicker.

Fig. 15 shows how two fluorescent lamps are connected to reduce the stroboscopic effect (strobe flicker). Capacitor C causes the upper lamp to draw a leading current and thereby stagger the peaks of light output from the pair of lamps. Strobe flicker appears as a broken series of images when a moving object is viewed under a fluorescent lamp. Note the starting switch in Fig. 15. This switch is closed momentarily, and then released. As the switch opens, kickback voltages from the ballast reactors fire up the fluorescent lamps. The arc discharges are then self-sustaining until the AC input is switched off.

A mercury-vapor lamp (Fig. 16) consists of two electrodes at opposite ends of a glass tube containing vaporized mercury. These

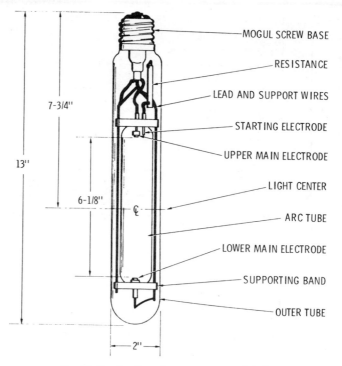

MOGUL SCREW BASE

RESISTANCE

LEAD AND SUPPORT WIRES

STARTING ELECTRODE

UPPER MAIN ELECTRODE

LIGHT CENTER

ARC TUBE

LOWER MAIN ELECTRODE

SUPPORTING BAND

OUTER TUBE

7-3/4"

13"

6-1/8"

2"

Fig. 16. Construction of a mercury-vapor lamp.

electrodes are made of coiled tungsten wire coated with chemical oxides for emission of electrons at a comparatively low temperature. The glass tube (arc tube) contains some argon gas to start the arc discharge before the mercury heats up and vaporizes. When voltage is applied to the lamp, an electric field is set up between the starting electrode and the upper main electrode. This causes electron emission from the surface of the main electrode. These moving electrons ionize the gas and start the arc discharge.

The sodium lamp shown in Fig. 13 is larger and longer than a mercury-vapor lamp. However, its power consumption is less—about 200 watts. It contains sodium which vaporizes after the lamp heats up. Some neon gas is placed in the lamp to provide easy starting and to heat the sodium. A thermostatic starting timer is used which applies power to the filaments at each end for ½ minute or longer. Then, the thermostat opens and full voltage is applied across the lamp which starts the arc discharge. A dozen or more sodium lamps are often operated in series from a special transformer that supplies a constant current of 6.6 amperes.

SUN LAMPS

Sun (sunlight) lamps, such as shown in Fig. 17, provide a combination of visible radiation and ultraviolet light. A tungsten filament is used which is connected in parallel with electrodes for a mercury-vapor

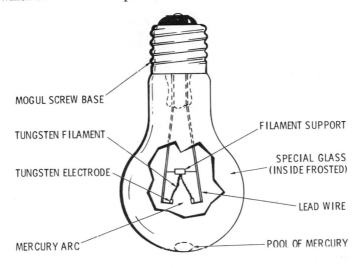

Fig. 17. Construction of a typical sun lamp.

arc. Some argon gas is also placed in the bulb. When voltage is applied to the lamp, the filament is heated to a sufficiently high temperature and it emits electrons. The moving electrons ionize the argon gas and an arc discharge is started. In turn, the mercury quickly vaporizes, and the resulting mercury arc draws a heavy current; the voltage drop across the filament decreases and the filament temperature drops several hundred degrees.

Sun lamps are powered by transformers with high leakage reactance, so that the current is limited. The transformer is constructed to serve as a ballast for the lamp. Electricians must make certain that correct voltage is supplied to the transformer, because the lamp is likely to

231

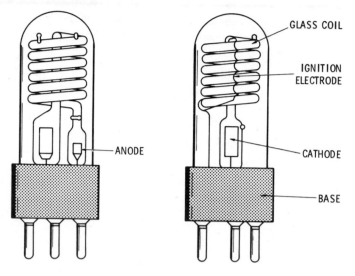

Fig. 18. High-voltage flash tube.

operate unsatisfactorily if the supply voltage is 5% too high or too low. The most popular sun lamp for home use is rated at 275 watts. More powerful ultraviolet lamps are used in hospitals. Since ultraviolet radiation can damage the retina of the eye, protective goggles must be worn.

The action of a transformer used to power a sun lamp can be compared with the action of an ideal transformer with an inductor connected in series with the primary circuit. In other words, leakage reactance occurs when all of the magnetic flux produced by the primary does not cut the secondary turns. When only a fraction of the primary flux passes through the secondary, it is the same as if the inactive part of the primary winding were connected externally in series with the primary of an ideal transformer. This inactive part of the primary winding develops the leakage reactance that limits current output from the secondary.

If we check the secondary voltage in the absence of a load, the primary draws a very small amount of current. In turn, there is only a very small voltage drop across the leakage reactance. On the other hand, if we check the secondary voltage under heavy load, the primary draws a large amount of current. In turn, there is a large

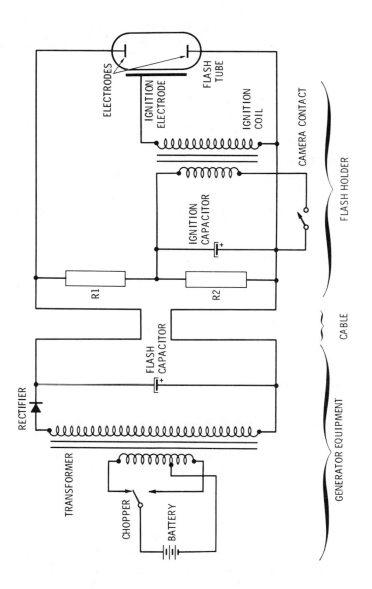

Fig. 19. Flash-tube circuit.

voltage drop across the leakage reactance. This is just another way of saying that the secondary voltage decreases rapidly with increasing loads when a transformer has high leakage reactance. This is why the amount of current that can be drawn from the secondary is limited by the value of the leakage reactance.

OZONE-PRODUCING LAMPS

Ozone is a form of oxygen that is formed in air by radiation of suitable wavelengths. Ozone-producing lamps are basically mercury-vapor lamps that produce radiation at a wavelength of 1849 angstroms. A typical lamp is rated at 4 watts and operates at 105 volts. This type of lamp is used as a deodorizer; ozone is a very active form of oxygen that reacts with common ill-smelling vapors and changes them into inoffensive chemical compounds. Ozone has a sharp and distinctive odor which is not unpleasant. Electricians who work in the vicinity of sparking electrical machinery are familiar with the odor of ozone.

FLASH TUBES

Flash tubes use straight or coiled sections of glass or quartz tubing, as shown in Fig. 18, and are filled with a gas such as xenon. When a voltage surge is applied to the tube, it produces a short flash of white light useful for high-speed photography. Operating voltages range from 800 to several thousand volts. Fig. 19 shows a flash-tube circuit. The system is battery-powered, with a vibrator to chop the DC supply into an AC voltage to energize the transformer. In turn, a high AC voltage is provided by the secondary winding. The rectifier is a "one-way" valve for electric current, which charges the flash capacitor in the polarity shown in the diagram.

Electrical energy is stored in the flash capacitor shown in Fig. 19, but does not produce an arc discharge until a high-voltage pulse is applied to the ignition electrode. Resistors R_1 and R_2 operate as a voltage divider, and the ignition capacitor is charged to a lower voltage than the flash capacitor. To trigger the flash tube, the camera contact is closed. In turn, a surge of current flows through the primary of the ignition coil. The stepped-up voltage from the secondary is applied to the ignition electrode, and the xenon gas ionizes, permitting the flash capacitor to produce a brief arc discharge in the flash tube.

A flash tube produces a brilliant light of up to 90,000,000 peak lumens which lasts from a thousandth to a millionth of a second. The life of a typical flash tube extends for several thousand operations. Simpler flash tubes with less light output are used in amateur equipment, as shown in Fig. 20. With suitable high-speed switching facilities,

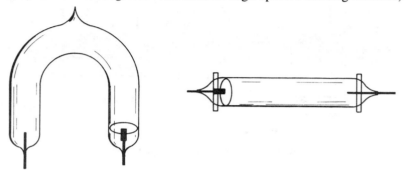

Fig. 20. Flash tubes used in amateur equipment.

flash tubes can be operated to produce a rapid series of light pulses of up to 30,000 per second. This method is used to "freeze" the motion of high-speed objects stroboscopically, thus providing a series of photos that show the details of the motion.

INFRARED LAMPS

The infrared range is below the visible range of radiant energy, as indicated in Table 1. Infrared lamps are used chiefly for heating and drying (see Fig. 2). They are often called heat lamps, and are rated from 125 to 1000 watts. Heat lamps usually operate at 115 volts. Tungsten filaments are used at a comparatively low temperature that produce much less light than an ordinary incandescent lamp. Most of the radiant energy is in the infrared region of wavelengths. Therefore, electricians describe this type of lamp as an "infrared heating" device. They are often used in radiation ovens. The lamps surround or partially surround the product that is to be dried or baked. Radiation ovens are best adapted for surface heating.

ELECTROLUMINESCENT PANELS

Fig. 21 illustrates the construction of electroluminescent panels. When a fluorescent material is embedded in the insulating dielectric between the plates of a capacitor, light can be produced by the electro-

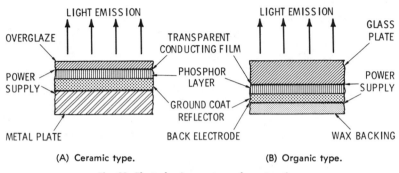

(A) Ceramic type.   (B) Organic type.

Fig. 21. Electroluminescent panel construction.

static field between the plates of the capacitor. If one of the plates is made of a transparent conducting substance, the light will be radiated into surrounding space. This type of light source is called an EL plate or panel.

EL panels are not lamps as ordinarily defined by electricians. However, they are similar to lamps in that they provide lighted surfaces that are useful in many practical applications. An EL panel operates in the range from 115 to 600 volts. Typical applications are in faces of electric clocks and cover panels for electric switches or outlets. They are also used in aircraft signs, such as "Fasten Seat Belts." This type of EL panel remains illuminated only while voltage is applied to the capacitor plates. AC is commonly used to energize ordinary EL panels.

Another type of EL panel is called an image-retaining panel, and is used in industrial plants. It employs a sulphide phosphor in a ceramic plate which is energized by DC voltage. This type of panel does not radiate light, however, until visible light, Xrays, or electron beams strike the panel. At any point on the panel, the radiated light has an intensity that depends on the intensity of the visible light, Xrays, or electron beams. Therefore, an image is formed which is retained as long as the DC voltage is applied. After the DC voltage is switched off, the image disappears. This type of panel is useful in Xray testing of manufactured products because the image can be inspected at leisure after the Xray beam is turned off.

SUMMARY

A unit of illumination is called a foot-candle. If the entire inner surface of a sphere is illuminated by an amount of 1 foot-candle, the total light output from the source is 1 lumen. The intensity of illumination falls off as the square of the distance from the source.

Only metal with a high melting point is used as a lamp filament. Tungsten melts at approximately 6120°F, and a typical tungsten lamp operates at 4800°F. The average lamp life is approximately 1000 hours because the tungsten filament gradually evaporates at this temperature. Bulb blackening is caused by slow tungsten evaporation. Most filaments are coiled in the form of a long helical spring which reduces the tendency of the gas to cool the filament.

Vapor discharge lamps produce light by means of the flow of electricity through gases. The general types of lamps are mercury, sodium, neon, neon glow, fluorescent, and ultraviolet germicidal. Vapor discharge lamps are generally classed as hot-filament, low-voltage, high-voltage, cold-cathode, and mercury-pool arc.

Infrared lamps, sometimes called heating lamps, are used for heating and drying, and are often used in radiation ovens. Tungsten filaments are used at a comparatively low temperature that produce much less light than an ordinary incandescent lamp.

Sun lamps are powered by transformers so that the current is limited. The transformer serves as a ballast which stabilizes the voltage. Tungsten filaments are used and are connected in parallel with electrodes for a mercury-vapor arc.

TEST QUESTIONS

1. What is the difference between various colors of light in terms of wavelength?
2. To what color is the human eye most sensitive?
3. State an application for an infrared lamp; for an ordinary tungsten lamp; for an ultraviolet lamp.

4. What is an instrument used to measure illumination called?
5. Why is tungsten a good material to use for lamp filaments?
6. Define a lumen.
7. Explain the inverse-square law.
8. Describe bulb blackening.
9. Why does a sudden surge of current flow when a tungsten lamp is switched on?
10. What do electricians mean by the "smashing point" of a tungsten lamp?
11. How are the various types of lamp bulbs identified? How are their diameters identified?
12. Describe a vapor-discharge lamp.
13. How does a fluorescent lamp operate?
14. Why is a ballast required with a vapor-discharge lamp?
15. Explain the operation of a sodium lamp.
16. How is a sun lamp constructed?
17. Describe an ozone-producing lamp.
18. How is a flash tube constructed?
19. Explain the circuit used to operate a flash tube.
20. Discuss an infrared lamp.
21. What is an electroluminescent panel?
22. State an application for an electroluminescent panel.
23. Explain an image-retaining panel.
24. State an application for an image-retaining panel.

CHAPTER 9

Lighting Calculations

Electricians are often concerned with lighting calculations to determine how much illumination is required in a given installation and to determine the necessary installation details. Two basic methods are used to calculate illumination requirements:

1. Point-by-point method.
2. Lumen method.

The final calculations are made in terms of *watts per square foot.*

POINT-BY-POINT METHOD

This method permits the calculation of the illumination at any given point by means of simple arithmetic. We consider the candle-power distribution of the light source and its position with respect to the given point. It is not necessarily the best method to use in any installation, although electricians prefer this method in certain lighting problems. The point-by-point method is an application of the inverse-square law. For the reader's convenience, an illustration of the inverse-square law is repeated in Fig. 1. It shows that the intensity of illumination produced by a point source of light decreases as the square

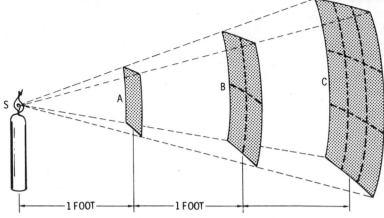

Fig. 1. Illustrating the inverse square law.

of the distance from the source. From a *candlepower distribution curve* of a reflector, the footcandles at any given point may be found from the formulas:

$$\text{Footcandles (perpendicular to the beam)} = \frac{\text{CP (candlepower)}}{\text{D}^2 \text{ (distance in feet)}}$$

$$\text{Footcandles (on horizontal plane)} = \frac{\text{CP}}{\text{D}^2} \times \text{cosine of angle } \theta$$

The meaning of *perpendicular to the beam* is shown in Fig. 1. The meaning of *angle* θ is shown in Fig. 2. The number of degrees in an angle is measured with a *protractor,* as shown in Fig. 3. Thus,

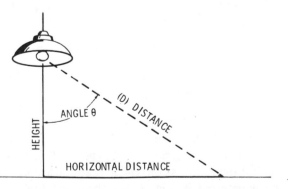

Fig. 2. Illustrating point-by-point method of lighting calculation.

240

the angle AOB is a 35° angle. With reference to Fig. 2, cosine θ is the height divided by the distance (D) of the right-angled triangle. Table 1 lists the cosines of angles from 0° to 90°. Therefore, when we wish to find the number of footcandles on the horizontal plane in

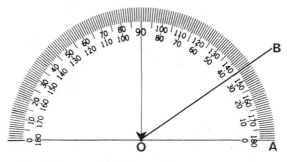

Fig. 3. How a protractor is used to measure an angle.

Fig. 2, we measure the angle θ. Then, we look in Table 1 and find the cosine of this angle. We must also measure the distance (D) and the height in Fig. 2. We multiply the candlepower of the light source by the cosine of angle θ and divide this product by the square of distance (D). The answer is the number of footcandles on the horizontal plane.

Electricians are sometimes concerned with the sine or the tangent of an angle. Values of sines and tangents are also given in Table 1. Note that the sine of angle θ in Fig. 2 is equal to distance (D) divided by the horizontal distance. The tangent of angle θ is equal to the horizontal distance divided by the height. Let us observe the vertical plane, perpendicular plane, and horizontal plane shown in Fig. 4. We know that the illumination on the perpendicular plane is found as follows:

$$\text{Footcandles (at point } d \text{ on the perp. plane)} = \frac{CP}{D^2}$$

We also know that the illumination on the horizontal plane is found as follows:

$$\text{Footcandles (at point } d \text{ on the horiz. plane)} = \frac{CP}{D^2} \times \text{cosine } \theta$$

241

Table 1. Sines, Cosines, and Tangents of Angles From 0° to 90°

Angle	Sin	Cos	Tan	Angle	Sin	Cos	Tan
0°	0.0000	1.0000	0.0000	46°	0.7193	0.6947	1.0355
1°	0.0175	0.9998	0.0175	47°	0.7314	0.6820	1.0724
2°	0.0349	0.9994	0.0349	48°	0.7431	0.6691	1.1106
3°	0.0523	0.9986	0.0524	49°	0.7547	0.6561	1.1504
4°	0.0698	0.9976	0.0699	50°	0.7660	0.6428	1.1918
5°	0.0872	0.9962	0.0875	51°	0.7771	0.6293	1.2349
6°	0.1045	0.9945	0.1051	52°	0.7880	0.6157	1.2799
7°	0.1219	0.9925	0.1228	53°	0.7986	0.6018	1.3270
8°	0.1392	0.9903	0.1405	54°	0.8090	0.5878	1.3764
9°	0.1564	0.9877	0.1584	55°	0.8192	0.5736	1.4281
10°	0.1736	0.9848	0.1763	56°	0.8290	0.5592	1.4826
11°	0.1908	0.9816	0.1944	57°	0.8387	0.5446	1.5399
12°	0.2079	0.9781	0.2126	58°	0.8480	0.5299	1.6003
13°	0.2250	0.9744	0.2309	59°	0.8572	0.5150	1.6643
14°	0.2419	0.9703	0.2493	60°	0.8660	0.5000	1.7321
15°	0.2588	0.9659	0.2679	61°	0.8746	0.4848	1.8040
16°	0.2756	0.9613	0.2867	62°	0.8829	0.4695	1.8807
17°	0.2924	0.9563	0.3057	63°	0.8910	0.4540	1.9626
18°	0.3090	0.9511	0.3249	64°	0.8988	0.4384	2.0503
19°	0.3256	0.9455	0.3443	65°	0.9063	0.4226	2.1445
20°	0.3420	0.9397	0.3640	66°	0.9135	0.4067	2.2460
21°	0.3584	0.9336	0.3839	67°	0.9205	0.3907	2.3559
22°	0.3746	0.9272	0.4040	68°	0.9272	0.3746	2.4751
23°	0.3907	0.9205	0.4245	69°	0.9336	0.3584	2.6051
24°	0.4067	0.9135	0.4452	70°	0.9397	0.3420	2.7475
25°	0.4226	0.9063	0.4663	71°	0.9455	0.3256	2.9042
26°	0.4384	0.8988	0.4877	72°	0.9511	0.3090	3.0777
27°	0.4540	0.8910	0.5095	73°	0.9563	0.2924	3.2709
28°	0.4695	0.8829	0.5317	74°	0.9613	0.2756	3.4874
29°	0.4848	0.8746	0.5543	75°	0.9659	0.2588	3.7321
30°	0.5000	0.8660	0.5774	76°	0.9703	0.2419	4.0108
31°	0.5150	0.8572	0.6009	77°	0.9744	0.2250	4.3315
32°	0.5299	0.8480	0.6249	78°	0.9781	0.2079	4.7046
33°	0.5446	0.8387	0.6494	79°	0.9816	0.1908	5.1446
34°	0.5592	0.8290	0.6745	80°	0.9848	0.1736	5.6713
35°	0.5736	0.8192	0.7002	81°	0.9877	0.1564	6.3138
36°	0.5878	0.8090	0.7265	82°	0.9903	0.1392	7.1154
37°	0.6018	0.7986	0.7536	83°	0.9925	0.1219	8.1443
38°	0.6157	0.7880	0.7813	84°	0.9945	0.1045	9.5144
39°	0.6293	0.7771	0.8098	85°	0.9962	0.0872	11.4300
40°	0.6428	0.7660	0.8391	86°	0.9976	0.0698	14.3010
41°	0.6561	0.7547	0.8693	87°	0.9986	0.0523	19.0810
42°	0.6691	0.7431	0.9004	88°	0.9994	0.0349	28.6360
43°	0.6820	0.7314	0.9325	89°	0.9998	0.0175	57.2900
44°	0.6947	0.7193	0.9657	90°	1.0000	0.0000	
45°	0.7071	0.7071	1.0000				

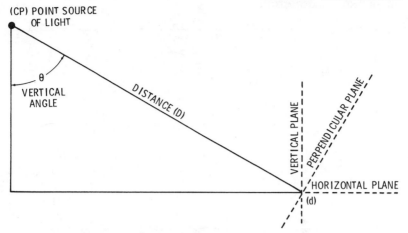

Fig. 4. Illustration of the vertical, perpendicular, and horizontal planes.

Next, the illumination on the vertical plane is found as follows:

$$\text{Footcandles (at point } d \text{ on the vert. plane)} = \frac{\text{CP}}{\text{D}^2} \times \text{sine } \theta$$

The ratio of illumination on the vertical plane to illumination on the horizontal plane can be found either from the two foregoing formulas, or the ratio can be found as follows:

$$\frac{\text{Footcandles at } d \text{ on vert. plane}}{\text{Footcandles at } d \text{ on horiz. plane}} = \frac{\text{CP}}{\text{D}^2} \times \text{tangent } \theta$$

When we use a CP value in an illumination formula, we must consider the candlepower distribution curve of the light source. Fig. 5 shows distribution curves for a tungsten lamp suspended vertically from a ceiling and for a tungsten lamp with a reflector. It is customary to draw a distribution curve to show the candlepower 10 feet from the light source and at all possible angles from the source. For example, curve *A* in Fig. 5A shows that a tungsten lamp has zero candlepower in the direction of its base. However, the lamp has 60 candlepower in a direction at right angles to its bulb. Again, the lamp has about 32 candlepower in a direction vertically below the bulb.

243

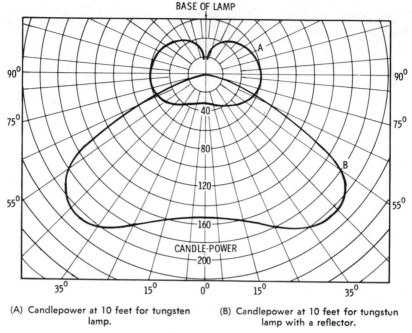

BASE OF LAMP

(A) Candlepower at 10 feet for tungsten lamp.

(B) Candlepower at 10 feet for tungstun lamp with a reflector.

Fig. 5. Distribution curve.

With reference to Fig. 5B, we observe that a reflector tends to concentrate the light in an area below the bulb. The lamp and reflector have 155 candlepower in a direction vertically below the bulb. However, there is zero candlepower in a direction at right angles to the bulb. At an angle of 45°, the lamp and reflector have about 205 candlepower. Distribution curves are very useful for comparison purposes. For example, curves A and B show that the addition of a reflector to the lamp makes the candlepower 155/32, or almost 5 times as great in a direction vertically below the bulb. This ratio holds true at 10 feet, at 20 feet, or at any distance.

LUMEN METHOD

The lumen method of lighting calculation is based on the *average* level of illumination desired over a given area. Since one footcandle is equal to one lumen per square foot, the total number of lumens which must be supplied to an area is found by multiplying this area

244

by the required level of illumination in footcandles. To find the total number of lumens which must be produced by the lamps, in order to supply the required number of lumens to the area, we must take light losses into account. These losses are due to absorption of light, both in the fixture and by the walls of the room, as well as the depreciation of the system due to gradual collection of dust, smoke, etc. The lumen method is used with the aid of suitable tables.

Reference is made to tables to determine the required level of illumination for various interiors. The type of lighting unit is selected. The mounting height and spacings are chosen with respect to the general dimensions of the room. In turn, the *room factor* or *room index* is based on the relation of these values. The *coefficient of utilization* is then found from the foregoing data, the efficiency and type of distribution of the lighting equipment, and the reflection factors of the walls and ceiling. Table 2 lists the percentages of light reflected from typical walls and ceilings. The coefficient of utilization is the percentage

Table 2. Percent of Light Reflected from Typical Walls and Ceilings

Surface	Class	Color	Light Reflected Per Cent
Paint		White	81
Paint	Light	Ivory	79
Paint		Cream	74
Caen Stone		Cream	69
Paint		Buff	63
Paint	Medium	Light Green	63
Paint		Light Gray	58
Caen Stone		Gray	56
Paint		Tan	48
Paint		Dark Gray	26
Paint		Olive Green	17
Paint	Dark	Light Oak	32
Paint		Dark Oak	13
Paint		Mahogany	8
Cement		Natural	25
Brick		Red	13

NOTE.—Each paint manufacturer's reflection values differ for similar colors, but the above table gives some idea of the colors in these three classifications and of the average reflecting qualities.

of the source light which actually reaches the given area. A depreciation factor must be estimated, and then the total number of lumens required from each lamp is found as follows:

Lumens required from each lamp =

$$\frac{\text{footcandles desired} \times \text{area per lamp in sq. ft.}}{\text{coefficient of utilization} \times \text{depreciation factor}}$$

or, when computing for lamps of various sizes, the equation becomes:

$$\text{Footcandles} = \frac{\text{lamp lumens} \times \text{coeff. of util.} \times \text{depreciation factor}}{\text{area per lamp in sq. ft.}}$$

In practice, it is necessary to select the nearest lamp size because lamps are rated in steps. It is helpful to use tables which summarize the results for the more common types of lighting units when using

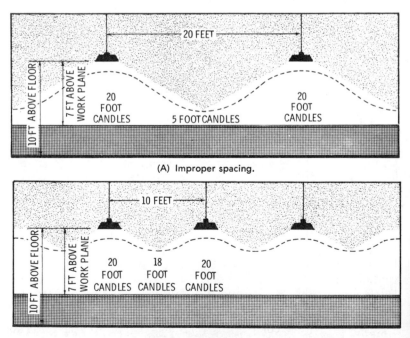

(A) Improper spacing.

(B) Normal spacing.

Fig. 6. Illustrating light fixture spacing.

the lumen method. This simplified method can be outlined for *interior lighting installation*:

1. Determine a suitable spacing of the lighting units (see Fig. 6).
2. Obtain the room factor (room index). This is explained below.
3. Determine the footcandles required, as explained below.
4. Determine the wattage necessary for the lamp or lamps, as explained below.
5. Calculate the necessary wire size, as explained in Chapter 2.

The correct spacing of lighting units to obtain practically uniform illumination throughout a room depends on several facts. Strictly speaking, the spacing required for uniform illumination depends on the height of the light source. Basically, the ceiling height limits the maximum permissible spacing. Note that the spacing of lighting units does not depend on the size or type of lamp that is used; the required spacing depends on the distribution curve of the reflector. For example, see Fig 5.

Rooms or halls may be divided by columns or beams into a definite number of sections or bays. Electricians usually try to make a symmetrical layout in accordance with the bays, partitions, and similar architectural features. Thus, the lighting units may be spaced as shown in Fig. 7. The arrangements in which each bay is treated as a separate unit have an additional advantage in that no change in lighting units will be required in case certain bays are partitioned later to make separate rooms.

The room factor is an appraisal; first from the standpoint of its general proportions; second, from the reflection factors of its walls and ceiling in order to determine what percentage of light is lost in traveling from the lamp to the illuminated area; third, from the type of lighting equipment to be used. In general, large rooms of average height use light more efficiently than small rooms of the same height because less of the total light from the source is absorbed by the walls. In using the simplified method, it is only necessary to note whether the room width is approximately equal to two, three, or four times the ceiling height.

Next, we note the color of the walls and ceiling. As listed in Table 2, the illumination in a room is dependent on the amount of light

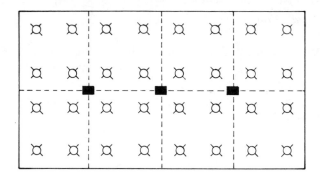

(A Four units per bay.

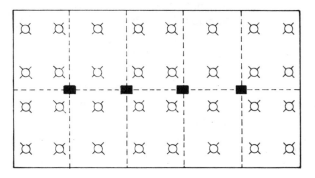

(B) Four-two system layout.

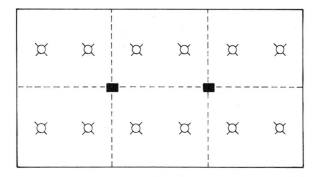

(C) Two unit per bay.

Fig. 7. Lighting

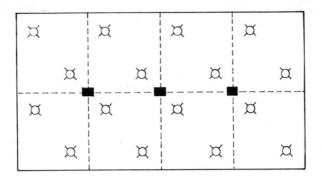

(D) Two units per bay staggered.

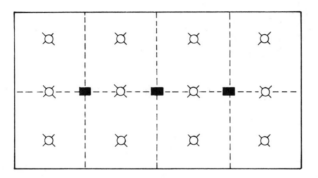

(E) Inter-spaced layout.

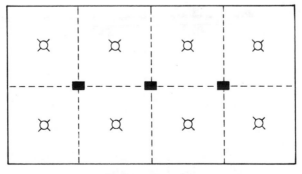

(F) One unit per bay.

arrangement.

249

reflected from the walls and ceiling. We use three classifications—light, medium. and dark. With the foregoing data, we can find the room factor from Table 3. We will use the room factor in combination

Table 3. Light Distribution of Various Types of Lighting Systems

Classification	Approximate light distribution	
	% Upward	% Downward
Direct	0-10	90-100
Semidirect	10-40	60-90
General diffusing	40-60	40-60
Semi-indirect	60-90	10-40
Indirect	90-100	0-10

with other data to find the lamp wattage that is required. At this point, let us briefly consider various types of lighting units.

LIGHTING EQUIPMENT

Lighting "fixtures" have changed considerably over the years with most of the present changes affecting the artistic forms. New diffusing and reflecting materials have been developed and there is a continuing

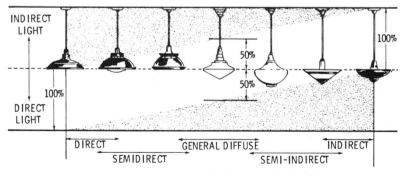

Fig. 8. Illustrating general classification of luminaires.

trend to "built-in" lighting systems. Thus, the electrician has a much wider range of lighting equipment to choose from than in the past. However, basic lighting calculations are changed very little by these developments. Lighting systems are grouped into four main types: (1) Direct; (2) Semidirect; (3) Semi-indirect; (4) Indirect. (See Fig. 9.)

Fig. 9. Valance lighting near the ceiling is a form of indirect lighting.

Direct lighting is defined as any system in which practically all of the light on the illuminated area is essentially downward and comes directly from the lighting units. Direct lighting systems range from spotlight and concentrating types of equipment, through the many types of bowl and dome reflectors, to extended light-source areas such as large glass panels and skylights. Fig. 8 illustrates the general classifications of luminaires. Open-type reflectors are most efficient, although it is difficult to provide high levels of illumination without glare unless considerable care is taken in locating and shielding this type of lighting unit. (See Fig. 10.)

Table 3 lists the light distributions of the various types of lighting systems illustrated in Fig. 8. To reduce the glare problem of open-type reflectors, louvered downlights with concentrating reflector or lens control may be used. This confines the light narrowly to the illuminated area with a minimum of light in the direction of the eyes. To obtain good distribution, careful location of equipment is required; harsh shadows and glaring reflections from shiny or polished surfaces should be avoided. Direct lighting from large-area sources with good diffusion

Fig. 10. Cornice lighting is a shielded form of direct lighting.

has some of the advantages of indirect lighting in that harsh shadows and both direct and reflected glare are minimized.

Semidirect lighting is provided by systems in which the predominant light on the illuminated area is obtained directly from the lighting units, with a considerable amount of light obtained by reflection from the ceiling. This is the type of lighting provided by opal or prismatic glass globes which surround the lamps. Lighting units of this type direct their light out at all angles, and are likely to be too bright for offices, schools, and similar installations unless the electrician uses oversize globes. However, it is sometimes practical to use parchment shades with the smaller globes to reduce brightness toward the eyes, and at the same time to redirect the light more efficiently to the illuminated area.

Semi-indirect lighting is defined as any system in which some light (such as 5% to 25%) is directed downward, with more than half of the light directed upward and reflected from the ceiling. The surface brightness of the bowl should not be greater than 500 foot-lamberts. Refer to Table 4 for the common units of brightness and their relations. The foot-lambert is equal to the incident footcandles multiplied by the reflection factor, assuming a diffusing illuminated area. The reflection factor is the ratio of light reflected by an illuminated area to the incident light. Opaque lighting units that have baffles or shielded

Table 4. Common Units of Brightness and Their Relations

1 Candle per square inch	= 452 foot-lamberts
	= 0.487 lamberts
	= 487 millilamberts
1 Foot-lambert =	1 lumen per square foot reflected or emitted
	= 0.00221 candles per square inch
	= 1.076 millilamberts
1 Lambert	= 1 lumen per square centimeter reflected or emitted
	= 1000 millilamberts
	= 929 foot-lamberts
	= 2.054 candles per square inch.
1 Millilambert	= 0.929 foot-lamberts
	= 0.002054 candles per square inch.

openings to redirect a small portion of the light to their undersurfaces for decorative effects are classed as indirect lighting units.

Indirect lighting has a soft and subdued appearance due to low brightness and absence of sharp shadows. Practically all of the light is diffusely reflected from large ceiling areas. Installations may consist of suspended or portable luminaires, or built-in concealed sources in the form of coves, ceiling coffers, column urns, or wall boxes. A fair uniformity of ceiling brightness is required. Practical problems may occur in the case of long, low-ceiling rooms, where large expanses of ceiling area are in the line of vision, if the installation produces more than 25 footcandles. These installation problems are less serious if the ceiling area is broken by crossbeams, or if ceiling valances are used to break up the flat area.

LEVEL OF ILLUMINATION

During past years there has been a trend toward higher levels of illumination in nearly all applications. (See Fig. 11.) Six classifications are recognized by the electrician, as follows:

1. Close and prolonged tasks with small details, low contrast, and high speed of operation. A level of at least 100 footcandles is recommended, which may be obtained by supplementary lighting.

2. When speed of operation is not a factor, the lighting level may be reduced to a minimum of 50 footcandles. The required level may be obtained by supplementary lighting.

253

Fig. 11. Supplementary lighting is widely used in the home.

3. For conventional industrial and commercial applications, a range of 20 to 50 footcandles is adequate. Economy may be realized in some installations by use of supplementary lighting.

4. Familiar tasks of comparatively brief duration, recreational activities, etc., require lighting levels from 10 to 20 footcandles. General lighting systems are commonly used in these applications.

5. With large objects, slow movement, and good contrast, 5 to 10 footcandles from a general lighting system is adequate.

6. Passageways with light traffic and no hazards may have lighting levels of less than 5 footcandles. The quality of illumination is not of great importance.

LAMP WATTAGE

After the electrician has selected a suitable illumination level, the necessary lamp wattage is then determined. Tables at back of book are used for this purpose. We locate the spacing between lighting units in the second column, and check the minimum mounting height given in the first column to make sure that the mounting height to be used is greater than the minimum height listed for the given spacing. Then, in the column headed *Room Factor,* we select a permissible room factor (A, B, C, D, or E) among those listed in accordance with the room factors specified in Table 5.

Following horizontally to the right of this line in the design-data table, we locate the column which contains the foot-candle value re-

254

Table 5. Room Factors

Proportions of Room	Color of Ceiling and Upper Sidewalls	Direct Lighting		Semi-Direct Lighting	Semi-Indirect Lighting	Indirect Lighting
		Distrib-uting	Concen-trating			
Width Approximat'ly Four or More Times Ceiling Height	Light	A	A	C	C	C
	Medium	A	A	C	D	D
	Dark	A	A	D	D	E
Width Approximately Twice Ceiling Height	Light	B	A	C	C	D
	Medium	B	B	D	D	D
	Dark	B	B	D	E	—
Width Approximately Equal to Ceiling Height	Light	C	B	D	D	D
	Medium	C	B	E	E	E
	Dark	C	B	E	—	—

commended for the installation At the head of the column will be found the size of Mazda lamp that is required. Although installation details vary in practice, these tables serve as a useful guide in almost any situation. When in doubt because of unusual cciling or sidewall surfaces, or because of unusual architectural features, it is good practice to use the next larger size of lamp.

WIRE CAPACITY

The basic facts concerning wire current-carrying capacity have been explained in Chapter 2. Note that the need for adequate wire capacity cannot be too strongly emphasized. Operation of lighting systems with poor wiring installations defeats the goal of better lighting. Moreover, operation may be uneconomical to the extent that losses suffered over a fiscal year are sufficient to pay for good wiring. Do-it-yourself electricians often install overloaded circuits; overload is not only a nuisance from the standpoint of fuse trouble or circuit-breaker tripping, but presents a hazard. Also, line-voltage fluctuation is inevitable and

255

will cause unsatisfactory operation of electrical devices in overloaded circuits.

Sometimes an attempt is made to obtain improved lighting in overloaded circuits by using lower-voltage lamps. This expedient provides improved illumination, but leads also to considerable variation in the illumination level when the circuit load changes. With light loading, lamps will be operated at over-voltage and will burn out early.

GENERAL WIRING DATA

The National Electrical Code is the basic guide for the wireman. Note, however, that the Code merely specifies wiring practice with regard to fire hazards, but with little consideration to economy of operation. For example, the size of wire used in a lighting installation may conform strictly to the Code and yet, because of the length of the circuit, have an excessive voltage drop. In turn, lamps will burn dimly, illumination will vary considerably with the load, and the user will be paying for excessive power consumption in line losses.

On either new or remodeling jobs, where the wattage has been determined, wiring specifications should be based on wattage for the next larger size of lamps. This is good foresight, inasmuch as the trend in the past has been to greater power consumption. Note that it costs about ⅓ more to double the capacity of an installation. In branch circuits for general illumination, the allowable *voltage drop* between panelboard and outlet is a maximum of 2 volts.

In some localities, wiring installations must meet specifications laid down in city, county, or state licenses with their associated installation rules. Public utilities usually have certain rules also, based on considerations of the best service that can be provided to the greatest number of customers. Any installation should be inspected by a qualified official to make certain that it is in accordance with NEC and local codes. In cities, such inspection is usually required by ordinances. Fire underwriters have inspectors in some localities to check wiring installations. Even if inspection is not mandatory in a given locality, it is good practice to request an inspection from the nearest underwriters' bureau. Installations in Federal and State buildings generally require inspection by authorized Federal or State personnel.

LOAD AND LENGTH OF RUN

For 15-amp circuits the initial load per circuit should not exceed 1000 watts, with No. 12 minimum wire size to be used where the length of run does not exceed 50 feet. No. 10 wire should be used for runs between 50 and 100 feet, and No. 8 wire for runs between 100 and 150 feet. For heavy-duty lamp circuits, with No. 8 wire and a 3000-watt load, runs may extend up to 50 feet; for runs from 50 to 100 feet, No. 6 wire should be used; No. 4 wire should be used for

Table 6. Wire Size Required (Computed for Maximum of 2-volt Drop on Two-wire 120-Volt Circuit)

Load per Circuit	Current 120-Volt Circuit	LENGTH OF RUN (Panel Box to Load Center)—Feet																	
Watts	Amps.	30	40	50	60	70	80	90	100	110	120	130	140	150	160	170	180	190	200
500	4.2	14	14	14	14	14	14	12	12	12	12	12	12	10	10	10	10	10	10
600	5.0	14	14	14	14	14	12	12	12	12	10	10	10	10	10	10	10	8	8
700	5.8	14	14	14	14	12	12	12	10	10	10	10	10	10	8	8	8	8	8
800	6.7	14	14	14	12	12	12	10	10	10	10	10	8	8	8	8	8	8	8
900	7.5	14	14	12	12	12	10	10	10	10	8	8	8	8	8	8	8	8	6
1000	8.3	14	14	12	12	10	10	10	10	10	8	8	8	8	8	8	6	6	6
1200	10.0	14	12	12	10	10	10	10	8	8	8	8	8	6	6	6	6	6	6
1400	11.7	14	12	10	10	10	8	8	8	8	8	6	6	6	6	6	6	6	6
1600	13.3	12	12	10	10	8	8	8	8	6	6	6	6	6	6	6	6	4	4
1800	15.0	12	10	10	10	8	8	8	6	6	6	6	6	6	4	4	4	4	4
2000	16.7	12	10	10	8	8	8	6	6	6	6	6	6	4	4	4	4	4	4
2200	18.3	12	10	10	8	8	8	6	6	6	6	6	4	4	4	4	4	4	2
2400	20.0	10	10	8	8	8	6	6	6	6	6	4	4	4	4	4	4	2	2
2600	21.7	10	10	8	8	6	6	6	6	4	4	4	4	4	4	4	4	2	2
2800	23.3	10	8	8	8	6	6	6	6	4	4	4	4	4	4	4	2	2	2
3000	25.0	10	8	8	6	6	6	6	6	4	4	4	4	4	4	2	2	2	2
3500	29.2	10	8	8	6	6	6	4	4	4	4	2	2	2	2	2	2	2	2
4000	33.3	8	8	6	6	6	4	4	4	4	2	2	2	2	2	2	1	1	1
4500	37.5	8	6	6	6	4	4	4	2	2	2	2	2	2	1	1	1	1	1

runs between 100 and 150 feet. It is recommended that panelboards be located so that the length of run does not exceed 100 feet whenever it is practical to do so. Table 6 lists required sizes for a maximum line drop of 2 volts.

PANELBOARDS

A panelboard is an electrical panel containing switches, fuses, or circuit breakers, housed in a metal cabinet. A panelboard may also be called a distribution panel. One spare circuit should be provided for each five circuits used in the initial installation. Concealed branch-circuit conduit should be large enough to add one additional circuit for every five or less circuits that it contains.

SERVICE AND FEEDERS

A line running from the source of supply to various branch lines is called a *feeder*. A feeder in a farm installation may be a *building feeder* from the main load panel to a branch-circuit panel in the same building or as a *service* to another building. Again, a feeder might be a *service feeder* running from a yard pole to a house or other building. The latter are often called *service drops,* but they operate as feeders. The *service* carries power into a building and to the load center where the power is then distributed to interior circuits.

The current-carrying capacity of service wiring and feeders should be such that there will be less than a 2-volt drop under full load. Provision should be made for a 50% increase in current-carrying capacity over the initial installation. Convenience outlets should not be connected to branch circuits which supply fixture outlets because the load that the user might place on a convenience outlet is somewhat unpredictable. A minimum wire size of No. 12 should be used for convenience outlets, and No. 10 should be used if the run exceeds 100 feet.

Electricians should have a National Electrical Code book at hand, and check requirements for outlets in various installations. Duplex outlets (double outlets) are specified in many situations. (See Fig. 12.) In office space, there should be one convenience outlet circuit for each

CONTACT SCREWS

Fig. 12. Duplex receptacle.

800 feet of floor area, and at least one duplex outlet for each 20 linear feet of wall. In manufacturing spaces, there should be one convenience outlet for each 1200 square feet of floor space, with at least one duplex outlet in each bay. In stores, there should be at least one convenience outlet in each supporting column or at least one floor outlet for each 400 square feet or fraction thereof in floor space. For windows, at least one outlet for each 5 linear feet of plate glass, with an additional floor outlet for each 50 square feet of platform area should be provided. Provision for signs should be made by installing a 1-inch conduit from the distribution panel to the front face of the building for each individual store space.

WATTS PER SQUARE FOOT

In determining illumination on a "watt-per-square-foot" basis, the floor area is computed from the outside dimensions of a building, the area, and the number of floors, not including open porches, garages connected with dwellings, nor unfinished spaces in basements or attics. Although the "watts-per-square-foot" basis is not exact, the illumination level may be found fairly well. One watt per square foot may produce from 3 to 10 footcandles, depending on room size, ceiling and wall color, type of lighting units, and method of lighting used. Therefore, any "watts-per-square-foot" load estimates should be based on required footcandles and also by the installation considerations.

One watt per square foot will provide about 5 footcandles in dead storage areas, locker rooms, inactive spaces, and so on. If a storage space is used as an active work area, two watts or more per square foot should be provided. *Two watts per square foot* will provide an illumination of 10 to 15 footcandles in industrial areas for familiar tasks of comparatively brief duration. In commercial areas, two watts per square foot will provide from 8 to 12 footcandles, when used with standard reflecting equipment.

Four watts per square foot will provide about 20 footcandles when somewhat greater illumination is required. *Six watts per square foot* provides about 30 footcandles, and is adequate for many conventional industrial and commercial applications. If indirect lighting is used, a higher allowance should be made. *Eight watts per square foot* will provide from 30 to 35 footcandles when somewhat greater illumination is required. Illumination requirements in excess of 35 footcandles are often met by use of supplementary lighting.

SUMMARY

There are basically two methods used to calculate illumination requirements; one is the point-by-point method, and the other is the lumen method. The point-by-point method permits calculation by means of simple arithmetic. The lumen method of lighting is calculated on the basis of the average level of illumination desired over a given area.

Lighting equipment is grouped into four main types; direct, semidirect, semi-indirect, and indirect. Direct lighting is defined as any system where practically all the light on the illuminated area is essentially downward and comes directly from the lighting fixture.

Semidirect lighting is provided by a system in which the light on the illuminated area is obtained directly from the lighting unit, with a certain amount of light obtained by reflection from the ceiling.

Semi-indirect is defined as any system in which 5% to 25% of the light is directed downward, with more than half of the light directed upward and reflected from the ceiling.

Indirect lighting has a soft and subdued appearance due to low brightness and the absence of sharp shadows. Practically all of the light is reflected from a large ceiling area.

A line running from the source of supply to various branch lines is called a feeder. A feeder could be a service feeder which is a line running from a yard pole to a house where the power is then distributed to interior circuits.

TEST QUESTIONS

1. What are the two basic methods used to calculate illumination requirements?
2. Describe the point-by-point method.
3. How do we find the cosine of an angle?
4. What do electricians mean by the perpendicular plane? Horizontal plane? Vertical plane?

5. Explain how a protractor is used.
6. Discuss a candlepower distribution curve.
7. Describe the lumen method of lighting calculation.
8. How many footcandles produce one lumen per square foot?
9. About how much light is reflected by medium-class paint?
10. Why must we observe a minimum spacing between lighting units in many installations?
11. Define the coefficient of utilization.
12. Give an outline of procedure for interior lighting installations.
13. What is the difference between direct, semidirect, general diffuse, and indirect luminaires?
14. Define a foot-lambert.
15. What is the recommended lighting level for close and prolonged tasks? For conventional industrial and commercial applications? For passageways?
16. What is meant by a room factor? How do we find a room factor?
17. State the maximum voltage drop that should be permitted between a panelboard and an outlet.
18. How many spare circuits should an electrician provide in a panelboard?
19. Describe service and feeder lines.
20. What is the maximum voltage drop that should be permitted in a service or feeder line?
21. What percentage increase in current-carrying capacity should be provided for future requirements in service and feeder installations?
22. Explain the "watts-per-square-foot" method of determining illumination requirements.
23. Why is the "watts-per-square-foot" method less exact than other methods?
24. About how many footcandles will one watt per square foot provide in dead storage areas?
25. Will eight watts per square foot provide eight times as many footcandles as one watt per square foot?

CHAPTER 10

House Wiring

House wiring must conform to regulations of the National Electrical Code as well as to the city electrical code applying to a particular locality. Methods of wiring that have been approved by the National Fire Protection Association include: open conductors, concealed knob-and-tube wiring, conduit wiring, surface metal raceways, armored cable, underfloor raceways, nonmetallic sheathed cable, electrical metallic tubing, cast-in-place raceways, and wireways and busways. Fig. 1 illustrates the meaning of these electrical terms.

PRACTICAL CONSIDERATIONS

The wiring of finished houses is not as easy as it might appear to the inexperienced electrician. No two houses are built alike, and no two wiremen would wire the same house in exactly the same manner. There are often various setbacks that make wiring difficult, such as parquet floors, double floors, clogged partitions, and other obstructions. By laying out the job and making a rough sketch, much labor and material will often be saved. In many cases, the only instruction given to the electrician is in the form of a plan showing the location and number of lights, from which he must figure out how to install them with the least amount of material and labor that will provide a good installation which will pass inspection.

A wireman must determine how many sockets are to be connected to each outlet; for example, a code might permit only 660 watts to each 2-wire circuit with 40 watts per socket. Base plugs are counted as sockets. After laying out the number of lights per circuit and the number of circuits, the center of distribution should then be found. If a large house has over four circuits, it is advisable to install a panelboard that will feed the various circuits. It should be installed at a central point.

Fig. 2 shows a plan for one floor of a dwelling wired with conduit. The numbers on the various outlets show the number of lamps that are supplied. This is an example of the loop wiring system. No branches are taken off between outlets. Four circuits are used, so that there will not be more than 10 lamps on any one circuit. Panelboards in loft buildings or other buildings that require 8 to 10 circuits to a floor should be distributed one to a floor. In a building covering a large area, it is often advisable to install two panels or centers to a floor, with two sets of feeders. Circuit lengths should be kept down to 100 feet or less. A careful layout of circuit centers will often save many feet of wiring.

A *riser* is a vertical wire connecting two or more floors in a wiring system. Fig. 3 shows a riser installation. Distributing centers or cutout cabinets should be installed near a partition which is located so that the running of risers is easy, and should be on an inside wall to guard against dampness. A cutout box is a fireproof cabinet with a hinged door that houses switches and fuses. If only one distributing point is to be used, it should be either in the basement or attic. In private houses, it is sometimes advisable to install only one panel for the entire house. This is good practice for a three-story house that does not require over 12 circuits.

Note that a panel is a switchboard unit made up of one or more sections for mounting switchgear apparatus. In some cases it is not advisable to install a panel, but to run the wires down to the basement and to the meter board where fuse blocks or circuit breakers for the various circuits are installed on the meter board.

SERVICE CONNECTIONS

The definition of a *service* is that portion of the supply conductors

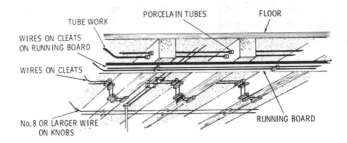

(A) Open conductors, knob-and-tube.

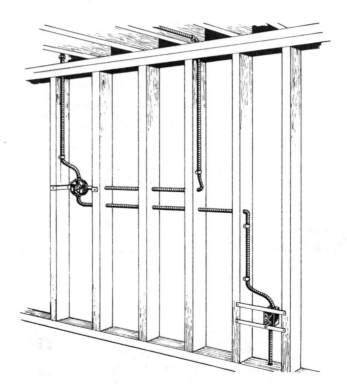

(B) Concealed wiring.

Fig. 1. Various house

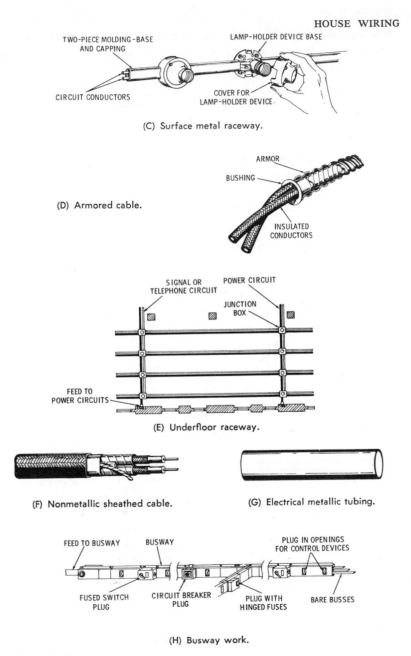

(C) Surface metal raceway.

(D) Armored cable.

(E) Underfloor raceway.

(F) Nonmetallic sheathed cable.

(G) Electrical metallic tubing.

(H) Busway work.

wiring methods.

265

which extends from the street main or duct or transformers, to the service switch, switches, or switchboard of the building supply. There are various methods of making a service entrance into a building, and they may be classified as tube or conduit service entrances. Fig. 4 shows a basic tube service entrance for a single wire. Fig. 5 shows basic conduit service entrances. The essential requirements for a tube service entrance are:

1. A dead-end insulator to carry the strain.
2. A connecting loop projecting downward to form a "drip."

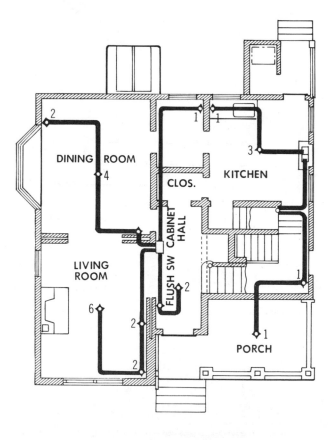

Fig. 2. A wiring plan for one floor dwelling.

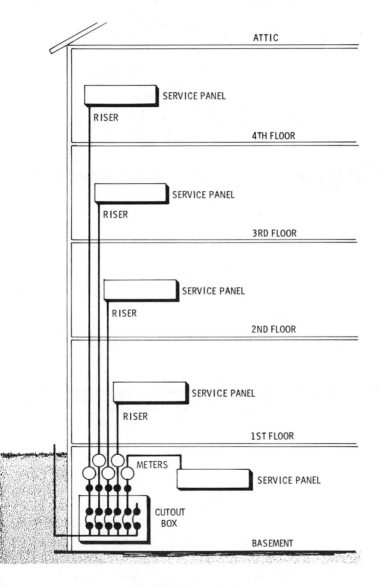

Fig. 3. Illustration of risers.

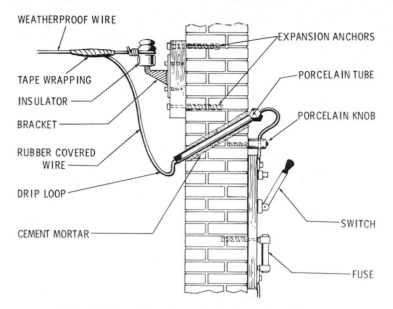

Fig. 4. Tube service entrance for single wire installation.

3. Porcelain-tube insulation where the wire passes through the wall.

4. Porcelain-knob insulation to keep the wire away from the inside wall and to prevent strain on the switch connection.

The conduit overhead service entrance is the most satisfactory method of running service wires from overhead wires to a building. If it should become necessary to install new wires, the change is more easily made than with tube service.

CUTOUT SWITCH

With reference to Fig. 5, rigid conduit is used from the cutout switch to a point at least 8 feet above the ground. The wires enter the conduit through a fitting called a *service cap* to protect the wires at the entrance point and to prevent water from entering the conduit. Pierce house racks and insulators are often used as illustrated in Fig. 6. A Pierce wire-holder insulator may be used, as shown in Fig. 7.

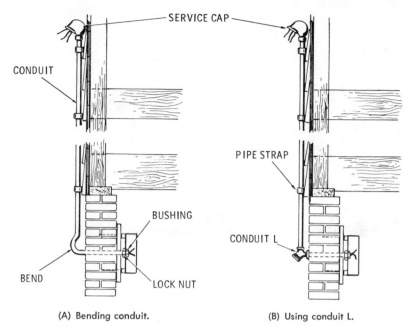

SERVICE CAP

CONDUIT

PIPE STRAP

BUSHING

CONDUIT L

BEND

LOCK NUT

(A) Bending conduit.   (B) Using conduit L.

Fig. 5. Service entrance cap and installation.

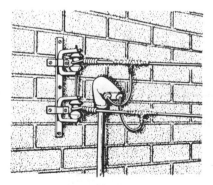

Fig. 6. Illustrating the Pierce house rack.

To install a conduit service entrance, a hole is drilled or bored through the wall to pass the conduit. The conduit is then bent so that the end passing through the wall extends ⅜ inch inside the main switch cabinet. Instead of bending the conduit, an approved conduit L fitting (not a common pipe elbow) may be used as shown in Fig. 8.

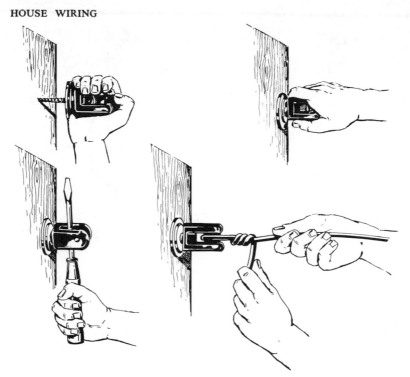

Fig. 7. Method of installing Pierce wire holder insulator.

The end of the conduit is secured to the switch box by a locknut and bushing. The locknut is screwed on the conduit before it enters the switch box. The bushing protects the wires where they leave the pipe, and should be tightened with a pair of gas pliers or combination pliers. The locknut is then tightened against the wall of the cabinet to hold the conduit securely in the switch box.

That portion of the conduit which is on the outside of the building is held in place by pipe straps, which in turn are fastened with screws. The L condulet must be of the weatherproof type. Fig. 9 shows one type of L condulet. This fitting is made weatherproof by placing a rubber gasket between the body of the fitting and the cover.

RESIDENTIAL UNDERGROUND SERVICE

There is a considerable trend to underground service in densely populated areas. An underground service lateral (wire or cable

270

MATERIAL

ITEM	DESCRIPTION
1	SERVICE TERMINATION ENCLOSURE (COMBINATION METER SOCKET PANEL ILLUSTRATED)
2	METER, WATTHOUR, SINGLE PHASE, SOCKET TYPE
3	CONDUIT, 1 1/2 MIN. ID FOR NO. 2 AL. SERVICE CABLE, MATERIAL PER CODE REQM'T (SEE NOTE)
4	SERVICE CABLE, 600 V, DIRECT BURIED, NO. 2 AL. MIN

ITEMS TO BE SUPPLIED BY POWER COMPANY

extending in a horizontal direction) is installed, owned, and maintained by the public utility. It is run to the residential termination facility, which is usually the kwh meter or meter enclosure, as depicted in Fig. 10. Meters are ordinarily located within 36 inches of wall nearest to the street or easement where the public utility's distribution facilities are located. The meter is mounted from 48 to 75 inches above final grade level. If there is likelihood of damage, meters must be adequately protected. Larger conduit is now required than formerly, and public utilities are specifying 1-½ inch inside diameter conduit for the service entrance. Note that aluminum conduit cannot be installed below ground level. The service conductors are usually No. 2 aluminum wire, and the service is customarily 3-wire 120/240 or 120/208 volt single-phase 60 Hz. The third wire is a grounded neutral conductor.

FEEDERS AND MAINS

In making a feeder layout for a large building, the current requirements on each floor should be noted. The best plan is to provide a feeder for every floor, especially in large installations. In smaller installations, one or two feeders may be all that are required. Feeders for motors should be independent of lighting feeders. In calculating sizes, feeders requiring over 2-inch conduit should not be used. It is better to subdivide feeders, especially if there are many bends or offsets, because 2-inch conduit is about the largest that can be economically handled.

Feeders should radiate from a distributing panel, with a properly rated switch and circuit breaker for each feeder. If the wiring system

271

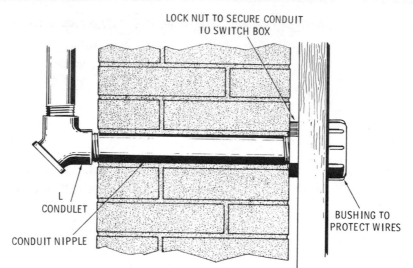

Fig. 8. Fitting used in securing conduit to switch box.

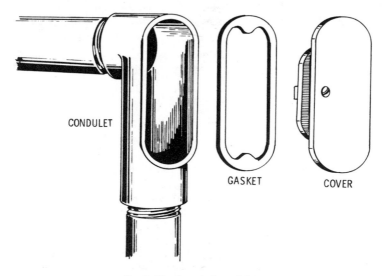

Fig. 9. Weatherproof conduit L.

is such that auxiliary power is taken from a local lighting company, it is a good plan to have each circuit controlled by a double-throw switch, so that in case of overload, any circuit can be fed from the

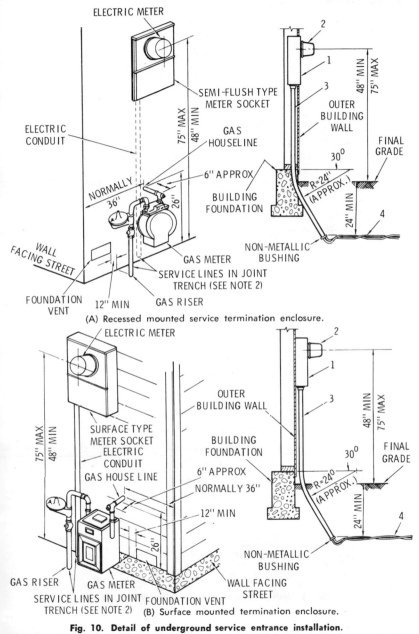

ELECTRIC METER

SEMI-FLUSH TYPE
METER SOCKET

ELECTRIC
CONDUIT

GAS
HOUSELINE

75" MAX
48" MIN

2

1

3

OUTER
BUILDING
WALL

48" MIN
75" MAX

FINAL
GRADE

NORMALLY
36"

6" APPROX

26"

30°

BUILDING
FOUNDATION

R=24"
(APPROX.)

WALL
FACING STREET

24" MIN

4

GAS METER

NON-METALLIC
BUSHING

FOUNDATION
VENT

12" MIN

SERVICE LINES IN JOINT
TRENCH (SEE NOTE 2)

GAS RISER

(A) Recessed mounted service termination enclosure.

ELECTRIC METER

OUTER
BUILDING WALL

2

1

SURFACE TYPE
METER SOCKET
ELECTRIC
CONDUIT
GAS HOUSE LINE

BUILDING
FOUNDATION

3

75" MAX
48" MIN

48" MIN
75" MAX

6" APPROX

FINAL
GRADE

NORMALLY 36"

30°

12" MIN

R=24"
(APPROX.)

26"

24" MIN

4

GAS RISER

GAS METER

NON-METALLIC
BUSHING

SERVICE LINES IN JOINT
TRENCH (SEE NOTE 2)

FOUNDATION VENT

WALL FACING
STREET

(B) Surface mounted termination enclosure.

Fig. 10. Detail of underground service entrance installation.

273

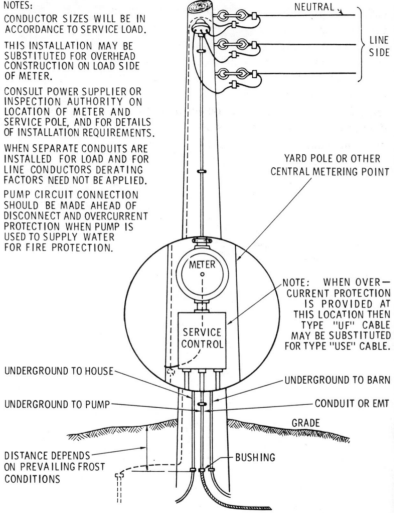

NOTES:
CONDUCTOR SIZES WILL BE IN ACCORDANCE TO SERVICE LOAD.

THIS INSTALLATION MAY BE SUBSTITUTED FOR OVERHEAD CONSTRUCTION ON LOAD SIDE OF METER.

CONSULT POWER SUPPLIER OR INSPECTION AUTHORITY ON LOCATION OF METER AND SERVICE POLE, AND FOR DETAILS OF INSTALLATION REQUIREMENTS.

WHEN SEPARATE CONDUITS ARE INSTALLED FOR LOAD AND FOR LINE CONDUCTORS DERATING FACTORS NEED NOT BE APPLIED.

PUMP CIRCUIT CONNECTION SHOULD BE MADE AHEAD OF DISCONNECT AND OVERCURRENT PROTECTION WHEN PUMP IS USED TO SUPPLY WATER FOR FIRE PROTECTION.

NEUTRAL
LINE SIDE

YARD POLE OR OTHER CENTRAL METERING POINT

METER

NOTE: WHEN OVER-CURRENT PROTECTION IS PROVIDED AT THIS LOCATION THEN TYPE "UF" CABLE MAY BE SUBSTITUTED FOR TYPE "USE" CABLE.

SERVICE CONTROL

UNDERGROUND TO HOUSE
UNDERGROUND TO PUMP
UNDERGROUND TO BARN
CONDUIT OR EMT
GRADE
DISTANCE DEPENDS ON PREVAILING FROST CONDITIONS
BUSHING

ALTERNATE METERING DRAWING SHOWING LOAD SIDE CONDUCTORS CARRIED UNDERGROUND TO BUILDINGS SERVED.

Fig. 11. Underground service entrance from a yard pole.

illuminating company's mains. It is advisable to install feeders and mains in conduit, even though the circuit wires are not. Feeders must carry the greatest current, and it is important to have them well protected since they usually run up side walls.

Wiremen should always check the applicable codes concerning restrictions against open or molding work on brick walls. Good protection is required, and this is an additional reason for piping the mains and feeders. In laying out branch circuits, it is not good practice to operate wiring at the maximum power permitted by a code. For example, if a power level of 660 watts is permitted, only 500 watts might be used in the installation. Note that if the initial installation operates at maximum permissible power, no lamps could be added in the future without violating the code.

In a farm installation, an underground service entrance may be made from a yard service pole, as shown in Fig. 11. The user must supply materials for the installation, with the exception of the meter which is supplied by the power company. Overcurrent protection (overload protection) is provided by a device that interrupts or reduces current flow in case excessive current is demanded. A *cable* is a stranded conductor; it is more flexible than a solid conductor. A cable *sheath* is the protective covering (usually lead) for excluding moisture and providing mechanical protection. UF is an abbreviation for *underground feeder*. USE means *underground service entrance*. EMT means *electrical metallic tubing* (thin-wall conduit).

CONDUIT FITTINGS

Conduit fittings are the accessories necessary for the completion of a conduit system, such as boxes, bushings, and access fittings. A conduit box is a metal box adapted for connection to conduit and is used to simplify wiring procedure or for mounting electrical devices. An access fitting is used to permit working with conductors elsewhere than at an outlet in a concealed or enclosed type of wiring system. Conduit differs from electrical metallic tubing in that conduit is comparatively heavy and uses threaded fittings.

A conduit fitting differs from an ordinary pipe fitting in that it has an opening with a removable cover, as shown in Fig. 12. This opening permits pulling wires through, as seen in Fig. 12A. A conduit fitting is used accordingly as a *pull box* and as a fitting to join two lengths of conduit. Various types of covers are used with conduit fittings. For

275

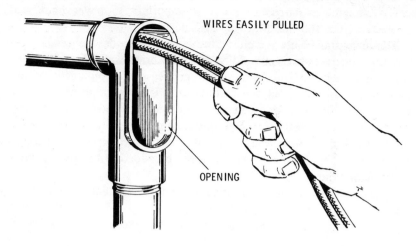

(A) Cover removed.

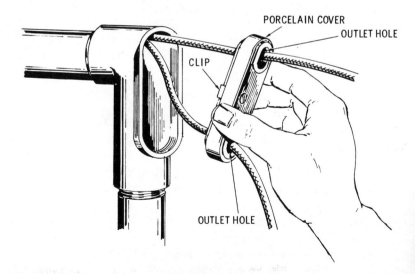

(B) Elbow with two-wire porcelain cover.

Fig. 12. Conduit elbow with removable cover.

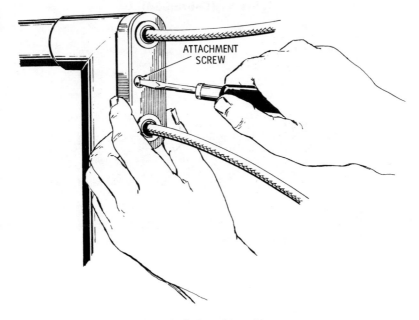

Fig. 13. Method of attaching elbow cover.

example, the type of cover shown in Fig. 12B provides an outlet for a branch circuit. Covers are fastened with machine screws, as shown in Fig. 13.

Table 1. Standard Electrical Symbols Used in Architectural Drawings

Ceiling Wall

		GENERAL OUTLETS		SWITCH OUTLETS
◯	⊸◯	Outlet.	S	Single-pole switch.
Ⓑ	⊸Ⓑ	Blanked outlet.	S_2	Double-pole switch.
Ⓓ		Drop cord.	S_3	Three-way switch.
Ⓔ	⊸Ⓔ	Electrical outlet—for use only when circle used alone might be confused with columns, plumbing symbols, etc.	S_4	Four-way switch.
			S_D	Automatic door switch.
			S_E	Electrolier switch.
			S_K	Key-operated switch.

277

Table 1. (Continued)

(F) –(F)	Fan outlet.	
(J) –(J)	Junction box.	
(L) –(L)	Lampholder.	
(L)$_{PS}$ –(L)$_{PS}$	Lampholder with pull switch.	
(S) –(S)	Pull switch.	
(V) –(V)	Outlet for vapor-discharge lamp.	
(X) –(X)	Exit light outlet.	
(C) –(C)	Clock outlet (specify voltage).	

S_P Switch and pilot lamp.

S_{CB} Circuit breaker.

S_{WCB} Weatherproof circuit breaker.

S_{MC} Momentary contact switch.

S_{RC} Remote control switch.

S_{WP} Weatherproof switch.

S_F Fused switch.

S_{WF} Weatherproof fused switch.

CONVENIENCE OUTLETS

⊖ Duplex convenience outlet.

⊖$_{1,3}$ Convenience outlet other than duplex. 1=single, 3 = triplex, etc.

⊖$_{WP}$ Weatherproof convenience outlet.

⊖$_R$ Range outlet.

⊖$_S$ Switch and convenience outlet.

⊖–[R] Radio and convenience outlet.

▲ Special-purpose outlet (description in specification).

⊙ Floor outlet.

SPECIAL OUTLETS.

○$_{a, b, c, etc.}$
⊖$_{a, b, c, etc.}$
$S_{a, b, c, etc.}$

Any standard symbol as given above with the addition of a lower case subscript letter may be used to designate some special variation of standard equipment of particular interest in a specific set of architectural plans.

When used they must be listed in the Key of Symbols on each drawing and if necessary further described in the specifications.

PANELS, CIRCUITS AND MISCELLANEOUS

Lighting panel.
Power panel.
Branch circuit—concealed in ceiling or wall.

Table 1. (Continued)

---- Branch circuit—concealed in floor.

----- Branch circuit—exposed.

→→ Home run to panelboard. Indicate number of circuits by number of arrows. NOTE: Any circuit without further designation indicates a two-wire circuit. For a greater number of wires indicate as follows:

---///--- (3 wires)

---////--- (4 wires), etc.

── Feeders. NOTE: Use heavy lines and designate by number corresponding to listing in feeder schedule.

Under-floor duct and junction box—Triple system. NOTE: For double or single systems eliminate one or two lines. This symbol is equally adaptable to auxiliary system layouts.

(G) Generator.

(M) Motor.

(I) Instrument.

AUXILIARY SYSTEMS

Pushbutton.

Buzzer.

Bell.

Annunciator.

Outside telephone.

Interconnecting telephone.

Telephone switchboard.

FS Automatic fire-alarm device.

W Watchman's station.

W Watchman's central station.

H Horn.

N Nurses's signal plug.

M Maid's signal plug.

R Radio outlet.

SC Signal central station.

Interconnection box.

Battery.

---- Auxiliary system circuits. NOTE: Any line without further designation indicates a 2-wire system. For a greater number of wires designate with numerals in manner similar to ---

12-No. 18 W-¾" C, or designate by number corresponding to listing in schedule.

a, b, c Special auxiliary outlets. Subscript letters refer to notes on plans or detailed description in specifications.

(T) Bell-ringing transformer.

D Electric door opener.

F◯ Fire-alarm bell.

F Fire-alarm station.

City fire-alarm station.

FA Fire-alarm central station.

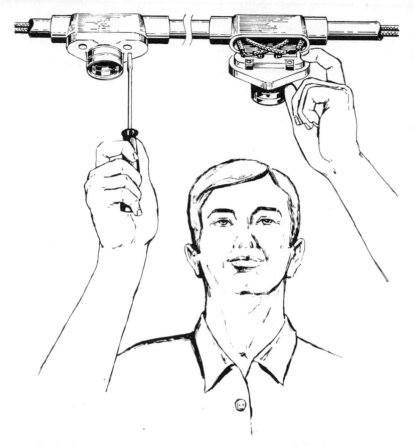

Fig. 14. Receptacle and cover showing method of connecting wires to base.

Covers with one hole are used for a drop cord to suspend a lamp from a ceiling. A *receptacle cover* provides connection for a light bulb, as shown in Fig. 14. This type of receptacle cover is called an outlet *coupling*. A *fixture cover* provides for connecting fixtures along a conduit run. These are suitable for installing fixtures of either the rigid or chain type.

Circuiting is a term used by practical electricians for planning the wiring used to connect lighting fixtures and switches to the source of electricity. Note that architectural drawings for new buildings do not show any electrical wiring in most cases. An architectural drawing merely indicates locations for outlets and receptables, switches, etc. Table 1 shows a list of standard electrical symbols used in architectural drawings.

The conduit elbow shown in Fig. 12 is a 90° type. Other types of elbows are also used in particular installations, as explained subsequently. Before an electrician can determine the types of conduit fittings that will be required in an installation, he must prepare a wiring diagram. This is a sketch showing each wire in an installation or part of an installation, with all the connections between the lamps and devices to the source of electricity. Note that *black* insulated wire

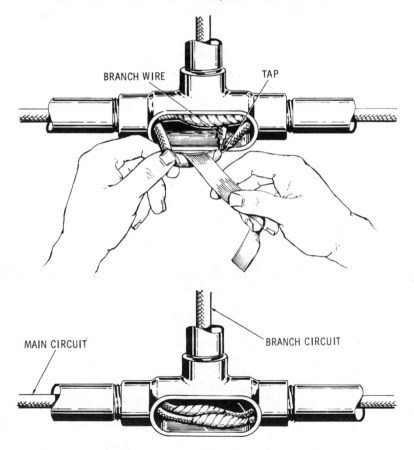

Fig. 15. Conduit T, showing method of tapping a branch circuit.

is used on one side of a run, and *white* insulated wire is used on the other side of the run. White insulation is used for return conductors, called the *grounded* or *neutral* wire.

Electricians call a black wire a "live" wire. There is voltage between a live wire and ground, but there is no voltage between a grounded wire and ground. Color coding of wires is very useful in practical work to avoid "getting wires crossed" when making splices in outlet couplings as shown in Fig. 14. When a pair of wires from each end of a conduit are to be spliced in an outlet coupling, we connect the two white wires together, and the two black wires together. Fig. 15 shows a *conduit T,* and how branch-circuit tap connections are made. Black wires are connected to black wires, and white wires are connected to white wires in tapping the main circuit.

Table 2. Two-Wire and Three-Wire Systems

Size of wire		Number of Wires in One Conduit								
		1	2	3	4	5	6	7	8	9
		Minimum Size of Conduit in Inches								
No.	14	½	½	½	¾	¾	1	1	1	1
	12	½	½	¾	¾	¾	1	1	1	1¼
	10	½	¾	¾	1	1	1	1¼	1¼	1¼
	8	½	¾	1	1	1	1¼	1¼	1¼	1¼
	6	½	1	1¼	1¼	1½	1½	2	2	2
	5	¾	1¼	1¼	1¼	1½	2	2	2	2
	4	¾	1¼	1¼	1½	2	2	2	2	2½
	3	¾	1¼	1¼	1½	2	2	2	2½	2½
	2	¾	1¼	1½	1½	2	2	2½	2½	2½
	1	¾	1½	1½	2	2	2½	2½	3	3
	0	1	1½	2	2	2½	2½	3	3	3
	00	1	2	2	2½	2½	3	3	3	3½
	000	1	2	2	2½	3	3	3	3½	3½
	0000	1¼	2	2½	2½	3	3	3½	3½	4
	200000	1¼	2	2½	2½	3	3	3½	3½	4
	225000	1¼	2½	2½	3	3	3½			
	250000	1¼	2½	2½	3	3	3½			
	300000	1¼	2½	3	3	3½	3½			
	350000	1¼	2½	3	3½	3½	4			
	400000	1¼	3	3	3½	4	4			
	450000	1½	3	3	3½	4	4½			

Table 2. (Continued)

Size of wire	Number of Wires in One Conduit								
	1	2	3	4	5	6	7	8	9
	Minimum Size of Conduit in Inches								
500000	1½	3	3	3½	4	4½			
550000	1½	3	3½	4	4½	5			
600000	2	3	3½	4	4½	5			
650000	2	3½	3½	4					
700000	2	3½	3½	4½					
750000	2	3½	3½	4½					
800000	2	3½	4	4½					
850000	2	3½	4	4½					
900000	2	3½	4	4½					
950000	2	4	4	5					
1000000	2	4	4	5					
1100000	2½	4	4½	6					
1200000	2½	4½	4½	6					
1250000	2½	4½	4½	6					
1300000	2½	4½	5	6					
1400000	2½	4½	5	6					
1500000	2½	4½	5	6					
1600000	2½	5	5	6					
1700000	3	5	5	6					
1750000	3	5	5	6					
1800000	3	5	6	6					
1900000	3	5	6						
2000000	3	5	6						

Conduit fittings are made in a large number of varieties, only a few of which can be illustrated in this chapter. In addition to the fittings shown previously, Fig. 16 shows some basic additional types. Note that 45° elbows are made in several forms, as follows:

1. With the opening facing the viewer, the 45° angle appears at the left-hand end.
2. In the reverse design, the 45° angle appears at the right-hand end.
3. With the opening facing down, and the trademark facing the viewer, the 45° angle appears at the right-hand end.
4. In the reverse design, the 45° angle appears at the left-hand end.

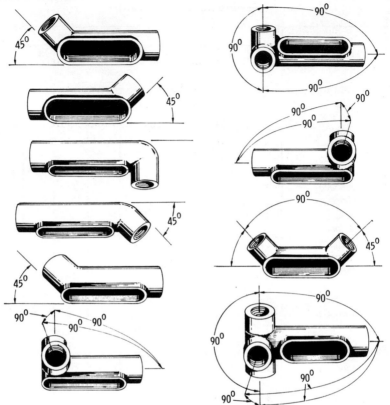

Fig. 16. Illustrating some of the many types of conduit fittings.

5. With the opening facing the viewer, a 45° angle appears at each end.

With reference to the 90° elbows, we note the following forms:

1. With the opening facing downward, and the trademark facing the viewer, the 90° angle appears at the right-hand end.
2. In a fitting with three projections, with the opening facing downward and the trademark facing the viewer, two 90° angles appear at the left-hand end.
3. With the trademark facing the viewer, the opening appears at a 45° angle between the 90° angles of the projections.

4. With the opening facing downward and the trademark facing the viewer, two 90° angles appear at the right-hand end.
5. In a fitting with four projections, with the opening facing the viewer, three 90° angles appear at the left-hand end.

Conduit fittings should be carefully chosen so that the conduit will be bent as little as possible. When selecting the size of conduit and fittings, applicable codes should be checked to make sure that the installation will pass inspection. The size of conduit depends on the size and number of wires that are used. Table 2 gives a general idea of the minimum size of conduit that will be used in various situations.

Table 3. Three-Conductor Convertible System

Size of Wires				Size Conduit Electrical Trade Size
two	14	and one	10	¾ Inch
"	12	"	8	¾ "
"	10	"	6	1 "
"	8	"	4	1 "
"	6	"	2	1¼ "
"	5	"	1	1¼ "
"	4	"	0	1½ "
"	3	"	00	1½ "
"	2	"	000	1½ "
"	1	"	0000	2 "
"	0	"	250000	2 "
"	00	"	350000	2½ "
"	000	"	400000	2½ "
"	0000	"	550000	3 "
"	250000	"	600000	3 "
"	300000	"	800000	3 "
"	400000	"	1000000	3½ "
"	500000	"	1250000	4 "
"	600000	"	1500000	4 "
"	700000	"	1750000	4½ "
"	800000	"	2000000	4½ "

Fig. 17 shows how wires appear in conduits. As noted above, these are merely typical examples and the electrician should be guided by applicable codes in selecting sizes of conduits for various sizes and

285

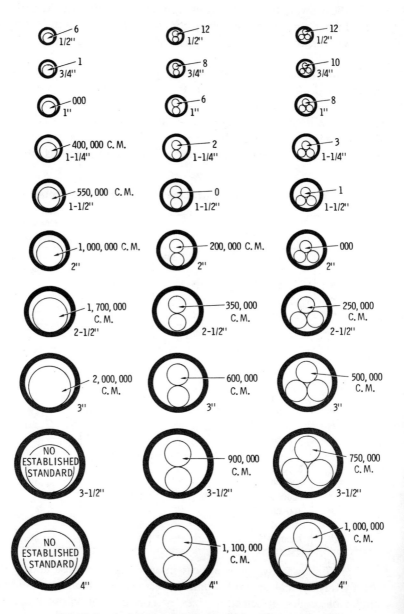

Fig. 17. Standard size conduit for installation of wire and cable.

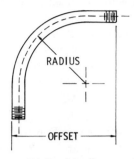

(A) Conduit ell.

(B) Conduit bending jaws or hickey.

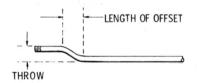

(C) An offset bend in rigid tubing.

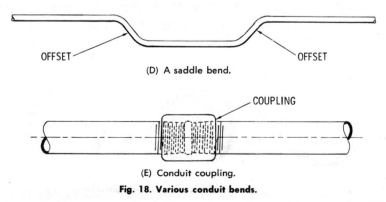

(D) A saddle bend.

(E) Conduit coupling.

Fig. 18. Various conduit bends.

numbers of wires. In some installations, one small wire may be run with a larger wire; in such a case, a suitable size of conduit must be used. Table 3 gives a general idea of the size of conduit that will be used.

Conduit elbows, usually called *ells,* have 90° bends and are measured in terms of their *offset* and *radius,* as shown in Fig. 18A. Ells are purchased ready-made because it is difficult to bend conduit to a short radius. However, if small conduit is bent in a fairly large radius, an ell can be made on the job. Various types of bench-mounted tools are used for bending conduit. Portable tools are preferred whenever they are practical. A portable bending tool is called a *hickey*. It consists of jaws (Fig. 18B), mounted on a long handle to provide good leverage.

The conduit to be bent is placed on the floor or ground, and the jaws of the hickey are placed around the conduit at a suitable point. Holding the conduit in place with his foot, the electrician then presses against the handle of the hickey and makes a small bend. Then, the bend is "sized up," and the position of the hickey is changed if necessary to complete the bend properly. Note that the inside radius of a bend should not be less than six times the diameter of the conduit. Otherwise, the conduit is likely to flatten, and it will be difficult or impossible to draw the wires through.

A hickey is used only to bend rigid-wall conduit; other types of tools are used to bend thin-wall conduit. An *offset* bend appears as shown in Fig. 18C. Two offsets produce a *saddle* bend, as shown in Fig. 18D. Other types of bends are also used, but the offset and the saddle bends are the basic types. Threaded *couplings* are used to connect lengths of conduit together, as shown in Fig. 18E. Practice is required to make accurate and neat bends with a hickey. When the length of throw and offset is comparatively small, two hickeys may be used separated by a suitable distance.

Conduit is *reamed* at both ends at the factory to remove sharp corners that could damage insulation on wires. When conduit is cut on the job, it should be reamed. A conical tool with cutting edges, called a *reamer,* is used for this purpose. Conduit is usually cut with a hacksaw or tubing cutter. The conduit should be held in a vise in order to make a workmanlike cut; the point of cutting should be carefully measured to avoid waste due to lengths that are a little too short for the installation. After the conduit is cut, it must be threaded using a stock and die as in plumbing work. An electrician uses a vise for cutting and threading conduit; the vise also serves for bending.

Conduit must always be grounded because an ungrounded run of conduit is both a shock hazard and a fire hazard. Conduit grounds use connections made with ground straps. A grounding conductor is run from the conduit to a water pipe on the supply side of the water meter. Grounding requirements are strictly regulated by electrical codes, which should be checked by the electrician to make sure that the installation will pass inspection.

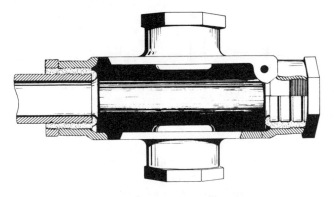

Fig. 19. Typical threadless conduit fitting.

Threadless conduit fittings are used with electrical metallic tubing, and may be used with rigid conduit in exposed runs. EMT cannot be threaded because its walls are quite thin. A threadless fitting (Fig. 19) has split sleeves which are tightened around the end of the conduit by locknuts. These fittings are made in many varieties and forms, and are commonly called *no-thread unilets*. For example, conduit fittings with 90° and 45° projections such as shown in Fig. 16 are also made in no-thread types.

EMT is lighter and easier to work with than rigid-wall conduit, and costs less. It can be bent easily with a suitable bender. Much time can be saved by using EMT in an installation if electrical codes permit its use. Conduit must be well supported and held firmly in place by suitable devices. Pipe clamps may be used in many installations. Where parallel runs of two or more conduits are required, a *hanger* support is commonly used. A *plate hanger* consists of a metal plate with a lip bent at right angles. Large holes are drilled in the plate to pass the conduit, and small holes are drilled in the lip for wood screws.

A plate hanger may be bolted instead of screwed to a wall or ceiling. Or, a plate hanger may be suspended by metal rods from a ceiling. Instead of a plate hanger, an electrician may use a *trapeze hanger* such as shown in Fig. 20. Many types of trapeze hangers can be ob-

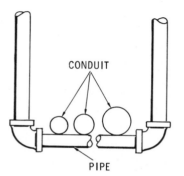

CONDUIT

PIPE

Fig. 20. One form of trapeze hanger.

tained from supply houses. A single conduit may be suspended by a *single-ring conduit hanger*. It consists of a ring-type support for the conduit, suspended from the ceiling by a metal rod, strap, or pipe.

Any run of conduit must be electrically continuous so that the entire run is well grounded. If a sharp edge or other defect in a conduit run cuts through the insulation of a live wire, the installation will be defective. Conduit is wired *after* the installation is completed. The wires are pushed or pulled (drawn) through the conduit, depending on its length. When wires are drawn through a long run of conduit, the procedure is called *fishing*. A *wire snake* is used for this purpose. The snake wire is fed into the conduit at one end by hand pressure as far as possible.

If more pressure must be applied in feeding the snake wire, a pair of pliers will give a firm grip. However, the snake can be fed only a few inches at a time if considerable pressure is necessary. After the snake comes out of the far end of the conduit, it is hooked to the wires, which are then drawn back through the conduit. Fishing from both ends of a long conduit is sometimes necessary. After a snake has been pushed as far as possible into one end of the conduit, another snake is pushed in from the other end. The hooks in the snake wires must then be caught. To do this, the electrician works with the short snake, while his helper shakes and works the long snake. After the hooks are caught, the electrician can finish pulling the long snake through.

Wires must be firmly attached to a snake hook, and it is good practice to tape the loops and turns. If a smooth tape covering is used, the attached wires are much less likely to jam in the conduit while being drawn. Drawing the wires requires one man to pull the snake and another man to feed the wires into the conduit. Wires must be prevented from kinking or crossing each other as they feed in. It is helpful to exert a steady pressure on the snake and to feed the wires smoothly instead of exerting a series of sudden jerks on the snake.

Wall switches are mounted above the fire blocks in a wall. Fire blocks are pieces of wood (such as $2'' \times 4''$) nailed horizontally (or somewhat diagonally) between studs or joists. They assist in the prevention of rapid spread fires and rise of hot gases inside the wall spaces. Fire blocks are ordinarily installed about 3½ feet above the floor level. In turn, the wall switches are mounted 4 feet or slightly less above the floor level. This is a convenient height, both for adults, and for children who are old enough to operate light switches.

HOW TO SPLICE WIRES

A two-wire cable or cord is stagger-spliced so that the splices do not overlap. To obtain a uniform stagger-splice that will ensure equal

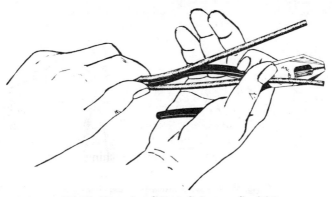

Fig. 21. Measuring distance between splice joints.

tension on both conductors when the splice is completed, the splice should be at least one plier length apart, as shown in Fig. 21.

The insulation is commonly removed by crushing the material with a pinching action by the plier handles, as illustrated in Fig. 22. The in-

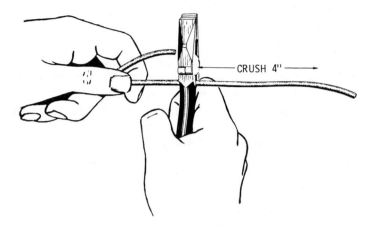

CRUSH 4"

Fig. 22. Crushing or pinching the insulation with plier handles.

sulation is crushed toward the end for a distance of about 4 inches, leaving about 2 inches of insulation uncrushed on the end of the conductor. This makes it easier to handle the wire in splicing them together.

The conductor is then skinned with a pair of pliers (see Fig. 23), or can sometimes be skinned with the fingers. A knife should never be used because the wire is easily nicked and weakened.

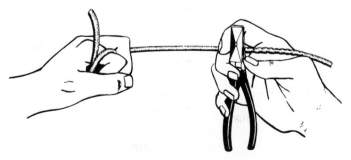

Fig. 23. Removing insulation from conductor.

A special type of electrician's pliers expedites insulation removal. These pliers are constructed from comparatively flat steel stock and have conventional wirecutting jaws at their tips. Just below the cutting jaws, a pair of blunt-edged spaces are provided for crimping jobs. Below the jaw hinge a pair of blades is provided, drilled with a succession of holes to pass wire sizes of 10, 12, 14, 16, 18, 20 and 22 AWG gauge. When the blades are pressed against an insulated wire at a suitable point, the insulation is cut without damage to the copper wire. Thus, the insulation is easily stripped from the end of the wire. Another useful feature of this type of electrician's pliers is provision of a special bolt cutter. Threaded holes in one arm of the pliers are opposed by a cutting blade on the other arm. The threads include 4-40, 5-40, 6-32, 8-32, 10-24, 10-32 sizes. When a bolt is screwed into a matching threaded hole, and the arms of the pliers are squeezed together, the bolt is cut cleanly without damage to the threads. This type of pliers is commonly provided with insulated handles.

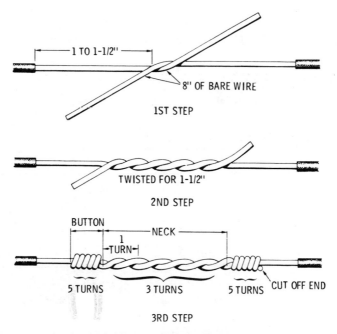

Fig. 24. Making the Western Union splice.

The most common splice for solid conductors is the Western Union splice. About eight inches of bare wire is required for this splice. The

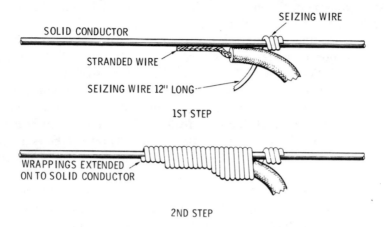

Fig. 25. Splicing stranded conductor to solid conductor.

wires are twisted together for about 1½ inches, after which the ends are bent at right angles to the axis of the wire (see Fig. 24). Each end

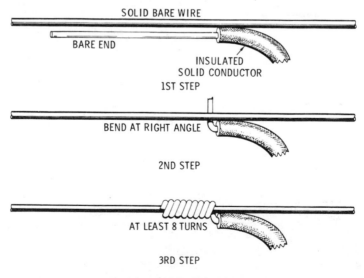

Fig. 26. Splicing solid conductor.

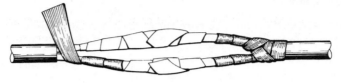

1ST STEP

2ND STEP

3RD STEP

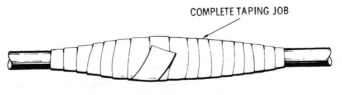

COMPLETE TAPING JOB

4TH STEP

Fig. 27. Applying rubber tape to splice.

is then wrapped around the wire for at least five close turns. The ends are then clipped off as close as possible.

A combination seizing wire splice is used to connect an insulated stranded conductor to a bare solid conductor. Strip about one inch of insulation from the end of the stranded wire and clean both wires. Lay the bared end of the stranded wire along the solid wire, as shown in Fig. 25. Take four turns with the seizing wire around the solid wire and continue the wrapping as shown in the diagram.

The commercial splice is occasionally used to connect an insulated solid conductor to a bare solid conductor, as shown in Fig. 26. Remove the insulation for about 6 inches. Bend the end at right angles to the bare wire and wrap for at least eight turns, as shown.

After a pair of wires have been spliced and soldered, rubber or plastic tape is wrapped around the splices, as shown in Fig. 27. The crotch

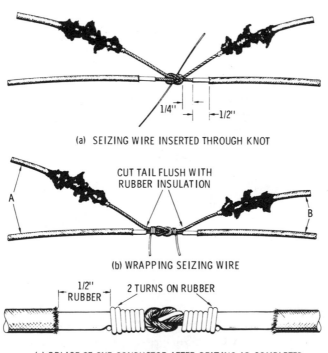

(a) SEIZING WIRE INSERTED THROUGH KNOT

1/4"   1/2"

CUT TAIL FLUSH WITH
RUBBER INSULATION

A

B

(b) WRAPPING SEIZING WIRE

1/2"
RUBBER   2 TURNS ON RUBBER

(c) SPLICE OF ONE CONDUCTOR AFTER SEIZING IS COMPLETED

Fig. 28. Splicing stranded wire.

laps are taped by one close turn and two additional turns around the wire. Finally, two layers of friction tape are wrapped over the rubber tape. Start this taping at the center of the splice and wrap to one end. cover the braid for one inch, and then bring the tape back to the other end. Cover the braid for one inch, and then wrap back, finishing in the center.

Stranded wires are commonly *seized* by knotting with a solid seizing conductor inserted through the knot. The seizing wire is then wrapped around the conductors as shown in Fig. 28. Finally, the surplus ends of the conductors and seizing wire are cut. The National Electrical Code requires that various splices shall be soldered, brazed, or welded. All splices, joints, or free ends must be covered with an insulation equivalent to that on the conductors.

SOLDERLESS CONNECTORS

Solderless connectors are permitted in many installations where there is no pull or force on the conductors, as inside an outlet box. Fig. 29A shows the solderless type of connector called a "wire nut". It is screwed over the bare ends of the wires to make the connection. A wire nut has an insulating shell, so that the connection does not need to be taped. Wire nuts are available in several different sizes. Fig. 29B shows the spring-type solderless connector. After it is screwed over the bare ends of the wires, the protruding "lever" is broken off. The completed connection must then be taped. Fig. 29C shows another type of solderless connector used on runs to buildings and feed lines for power. If a tap line exerts strain on an existing line, the type A connector is used. On the other hand, if there is little or no strain exerted, as in a service connection, the type B connector is used.

Solderless connectors can be used in all cases for copper-wire joints. However, certain solderless connectors may not be permitted for aluminum-wire joints. If a connector is suitable for use with aluminum wire, a statement to this effect will be printed on the connector package. Large-size connectors will be individually marked, as follows: A CU marking means that the connector can be used only with copper wires. An AL marking means that the connector can be used only with aluminum wires. A CU/AL

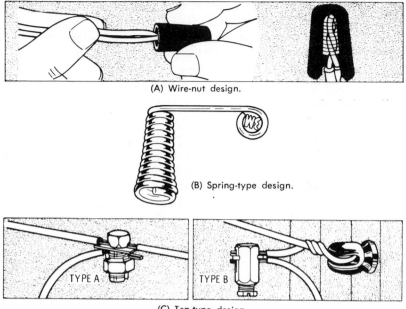

(A) Wire-nut design.

(B) Spring-type design.

(C) Tap-type design.

Fig. 29. Typical solderless connectors.

marking means that the connector may be used with either copper or aluminum wires; however, a copper wire cannot be joined to an aluminum wire in the same connector. If, in addition, the connector has a divider that prevents the copper and aluminum wires from touching each other, it can be used to join a copper wire to an aluminum wire. An unmarked large-size solderless connector can be used only with copper wires.

SUMMARY

House wiring must conform to the National Electrical Code regulations, as well as to all city regulations. There are many methods for wiring that are approved by the Code, which includes open conductor, concealed knob-and-tube wiring, conduit wiring, surface metal raceway, armored cable, underfloor raceway, nonmetallic sheathed cable, electrical metallic tubing, cast-in-place raceways, and wireways and busways.

A service connection is that portion of the supply conductor which extends from the street main or duct or transformer, to the service switch, switches, or switchboard of the building supply. The service entrance enters the conduit through a fitting called the service cap, which protects the wires at the entrance point and prevents water from entering the conduit.

Conduit fittings are made in a large number of varieties. Fittings should be carefully chosen so that the conduit will be bent as little as possible. Size of conduit will depend on the number of wires that are used. Conduit elbows are used which are ready-made and used where it is difficult to bend the conduit to a short radius.

Various types of bench-mounted tools are used for bending conduit. Portable tools are preferred whenever they are practical. A portable bending tool is called a hickey. A hickey is used only to bend rigid-wall conduit, other types of tools are used to bend thin-wall tubing.

When conduit is cut to length on the job, a tool called a reamer is used to ream out the sharp edges. These sharp edges can damage the insulation on the wires when pulled through the tubing.

TEST QUESTIONS

1. Do all localities have the same electrical code?
2. What is the first step to take in wiring a finished house?
3. Explain what is meant by a riser.
4. What is the definition of a service?
5. Name the two basic methods of installing a service entrance.
6. State the four essential requirements for a tube service entrance.
7. Describe an underground service entrance.
8. Can feeders for motors be used as lighting feeders?
9. How is a Pierce wire-holder insulator installed?
10. What are the features of a weatherproof condulet?
11. Explain what is meant by a pull box.
12. Define the term "circuiting" as used by practical electricians.
13. Why are black wires and white wires used in an installation?
14. Explain how a conduit T is used to tap off a branch circuit.

15. What is an ell?
16. Describe an offset bend. A saddle bend.
17. How is a hickey used?
18. Where are threaded couplings used?
19. Explain how a threadless conduit fitting holds the end of a conduit.
20. What is the purpose of a hanger support?
21. Why is a run of conduit always grounded?
22. Describe the process of fishing a wire or wires through conduit.
23. What difficulties may be encountered in using wire snakes?
24. Why is EMT preferred instead of rigid-wall conduit in situations where its use is permissible?
25. Explain how to make a ground connection that will pass inspection.

CHAPTER 11

Wiring With Armored Cable and Flexible Conduit

Armored cable (Fig. 1) can be used in dry locations unless local codes impose restrictions. Ordinary armored cable contains two wires,

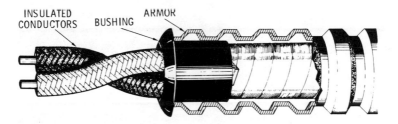

Fig. 1. Cutaway view of armor cable.

one with black and the other with white insulation. In a 3-wire cable, the third has red insulation. As in other types of conductors, armored cable is available in a wide range of wire sizes. While an installation with armored cable does not have the advantage of a conduit installation insofar as withdrawing old wires and inserting new wires, armored cable nevertheless has certain other advantages.

An armored cable is flexible and can be fished between floors and between partition walls. It is also less expensive than rigid conduit or flexible steel conduit. On the other hand, armored cable is more expensive than cleat wiring or knob-and-tube wiring. Armored cable is recommended over open wiring. Where subject to moisture, special types of armored cable are required by codes, such as *lead-covered armored cable*. The difference between lead-covered and ordinary armored cable is that the former has a continuous lead sheath over the wires which is covered, in turn, by flexible steel.

CUTTING ARMORED CABLE

Armored cable is supplied in rolls, much the same as other electrical cables. All types except ACL and ACV have an internal bonding strip of copper or aluminum in contact with the armor to provide good grounding. Type ACL cables have lead-covered conductors. Types ACV, AC, and ACT may be used in dry locations, except as noted in applicable codes. In order to remove the metal casing from any type of armored cable, a fine-toothed hacksaw (24 teeth to the inch) should be used, as shown in Fig. 2. A special armored-cable

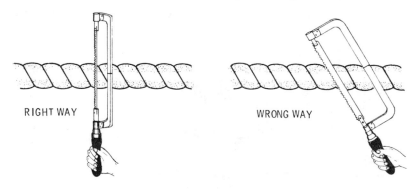

Fig. 2. Right and wrong way to cut armored cable.

tool may be used, if desired. Service-entrance cable is available in two types—ASE, with interlocked armor protection, and USE, underground service-entrance cable for direct burial in the ground. Al-

302

though constructional details vary, the method of removing the metal casing is basically the same.

Armor is cut diagonally. Do not cut entirely through the sheath, but make the cut deep enough so that the sheath will break when the cable is bent. Otherwise, you may damage the insulation or cut the wires. The saw may be used as shown in Fig. 3, supporting the cable on your knee, sawhorse, or any convenient object. After the armor sheath has been removed, the outer protecting braid is then removed. This can be done by making a slit one inch below the sheath about one inch long, and then pulling the outer braid off.

INSTALLATION OF ARMORED CABLE

Before the cable is installed, it should be examined at each end to see if the insulation might be punctured by any part of the sheath. This is an important precaution, because grounds and short-circuits are common in careless installations. To protect the insulation on the conductor, a bushing is inserted as seen in Fig. 4. Outlet boxes should be located and installed before the cable is installed. In turn, the electrician can "size up" the holes that will be necessary to be bored in order to run the cable. Holes should be bored through the floor beams at right angles to the run, so that the cable can be pulled through easily. Fig. 5 shows a run of armored cable.

After the holes have been bored for a run between two outlets, the cable may be pulled through. Then, the holes may be bored for a run between the next pair of outlets, and so on. In other words, it is not necessary to wait until all of the holes have been bored for the complete installation before starting the cable-pulling job. The cut end of the cable is first bushed, as shown in Fig. 6A. This is an important part of the job because the edges of the cut will have sharp edges projecting inward, as seen in Fig. 6B. When properly bushed, the wires will be completely protected from these sharp edges.

The cable is then pulled through the holes that have been bored and fastened with a clamp or connector to the outlet box. Knockouts in boxes are removed by hitting them with a hammer; this leaves an access hole for the cable. If a knockout has been removed and then is not used, it should be closed with a knockout closer. One type of connector is shown in Fig. 7A and the method of connecting it to the

303

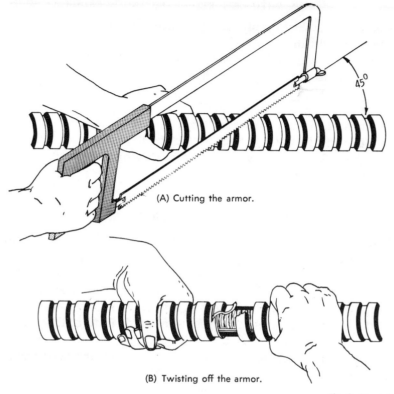

(A) Cutting the armor.

(B) Twisting off the armor.

Fig. 3. Removing

cable and box is seen in Fig. 7B. Another view of the box and completed connector is shown in Fig. 8.

After fastening the cable to the outlet box, it is pulled fairly tight and is cut off at the proper point to connect up with the other box. When cutting off the cable, make sure to cut it at the right length so that it will project into the box about six inches, or enough to make a workmanlike joint. In case a cable is cut too short, it is necessary to remove it and run another cable of correct length, because joints are not allowed between boxes. When a cable runs parallel with joists or studs, it is fastened by pipe clamps.

In wiring an old house, the procedure is much the same as in a connected knob-and-tube installation. The cable must be fished from one outlet to the next, and then fastened into the outlet boxes. It is good

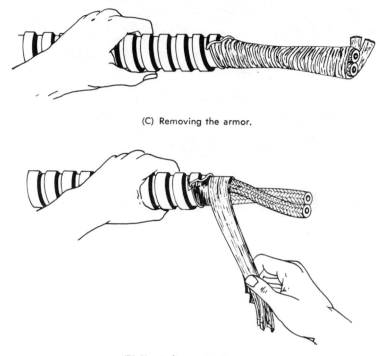

(C) Removing the armor.

(D) Unwinding protective cover.

the metal casing.

practice to fasten the armored cable to timbers with pipe straps wherever this may be possible. Approved straps or staples are used at 4½ foot intervals. At various points in an installation, the cable must be bent. All bends should be made as gradually as practical to avoid damaging and opening the armor. A cable should be securely fastened at all bends; this is particularly important when installing cable in the vicinity of machinery. Fig. 8 shows the right and wrong ways to bend armored cable.

FISHING ARMORED CABLES

In fishing cables, the electrician usually works from the attic over the wall outlet to be installed below. A hole is bored through the parti-

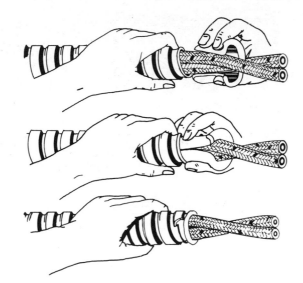

Fig. 4. Installing protective bushing.

CEILING OUTLET

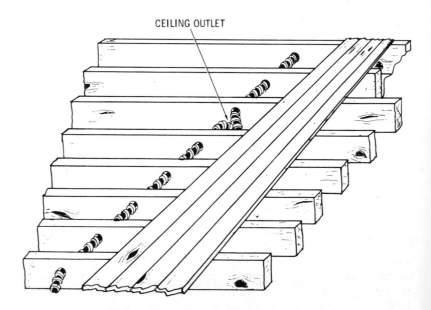

Fig. 5. A run of armored cable.

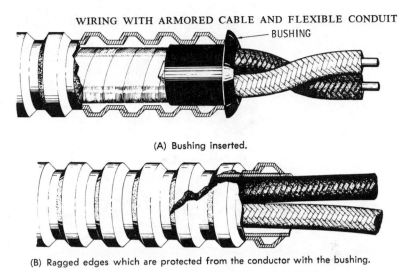

BUSHING

(A) Bushing inserted.

(B) Ragged edges which are protected from the conductor with the bushing.

Fig. 6. Illustrating how the bushing protects the conductor.

tion plate through which the armored cable will pass. Then, a chain is suspended and rattled in the space behind the wall opening for the outlet. Another man below fishes for the chain with a hooked wire. After drawing the chain through the opening, the armored cable is connected to the chain. Finally, the electrician draws the cable up through the wall to the attic.

Studs and braces in the wall space require suitable installation methods which are different depending on the location of wall outlets. (See Fig. 9.) If the chain from the attic strikes a brace, it may be practical to work from the nearest door. The work is concealed by removing the door stop and boring through the door casing and stud (or studs) over to the chain. If the distance is not too great, the chain can then be fished through the studs and out through the door casing. This completes the first step.

Next, a channel large enough to hold the cable is chiseled in the door casing down to a point that will clear the braces. Another hole is then bored through the door casing and studs over to the opening for the wall outlet. Then, the cable is fished from the opening to the door casing and attached to the chain for drawing up into the attic. After the door stop is replaced, the cable is completely concealed.

When a number of studs must be passed around braces, a hole is bored in the partition plate near the door casing, and the chain is fished

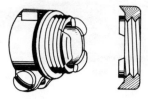

(A) Connector assembly.

(B) Installing connector bushing.

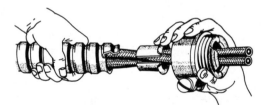

(C) Installing connector.

(D) Locking connector in position.

Fig. 7. Armored cable

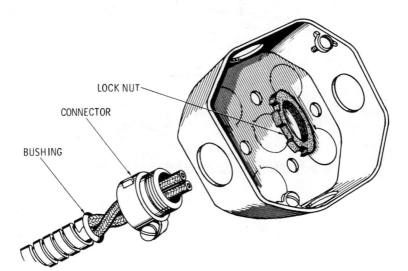

LOCK NUT

CONNECTOR

BUSHING

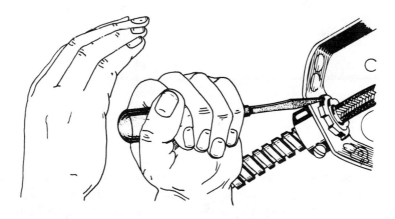

(E) Installing cable to junction box.

connector for junction box.

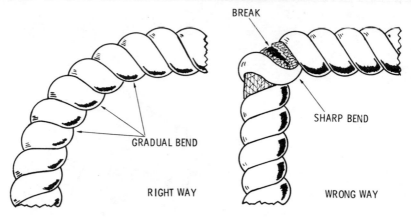

Fig. 8. Right and wrong way to bend armored cable.

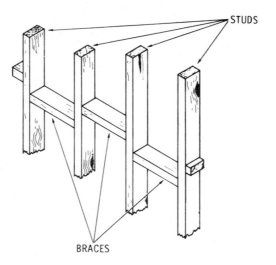

Fig. 9. Studs and braces require various installation methods.

as before. However, the channel in the door casing is chiseled down to the baseboard level so that the studs can be notched or bored after the baseboard is removed. This makes it easy to run the cable from the door casing to the opening for the wall outlet. If it is necessary to cross a door, the threshold is removed and a cable channel is chiseled in the floor below the threshold. After the threshold is replaced. the cable is

310

completely concealed. If there is no threshold, the door can be crossed at the top.

WHERE ARMORED CABLE CANNOT BE USED

The National Electrical Code forbids the installation of armored cable in theaters, except in some sound reproduction or communication circuits. Another exception is made in auditoriums with a seating capacity of less than 200 persons. Portable cables are permitted only where fixed wiring installations are impractical. Armored cables are forbidden in motion-picture studios, hazardous locations, wherever they might be exposed to fumes or vapors, on various cranes and hoists, in storage-battery rooms, or on elevators.

WIRING IN FLEXIBLE CONDUIT

A flexible conduit consists of a continous flexible steel tube which is made from convex and concave metal strips in a spiral winding upon each other so the concave surfaces interlock. (See Fig. 10.) Flexible

Fig. 10. Illustrating double-strip flexible conduit.

conduit has considerable strength and is supplied in lengths from 50 to 200 feet. Elbow fittings are not required because the conduit can be bent to almost any radius. There are fissures between the strips which provide some ventilation. This is an advantage in some applications and a disadvantage in others. Note that wires must be fished through flexible conduit.

The National Electrical Code forbids the use of flexible conduit in wet locations unless the conductors are of the lead-covered or equivalent type which has been approved for such locations. Flexible conduit is also prohibited in many hoistways, in storage-battery rooms, in most hazardous locations, and in locations where the rubber-covered conductors are exposed to oil, gasoline, or other chemicals which deteriorates rubber. Various other restrictions apply to installation of flexible conduit, both by national and local codes; therefore, the electrician should check before going ahead with a flexible-conduit installation.

Although flexible conduit is easy to work with, it is seldom desirable to install an entire wiring job with it. Electricians usually combine runs of flexible conduit with rigid conduit for extensions that are short and irregular. Occasional use of flexible conduit for short and irregular runs can save considerable pipe fitting. It can also be passed through joists and studs with comparatively little difficulty. Note that flexible conduit is also made in single-strip form, as shown in Fig. 11. Single-

Fig. 11. Illustrating single-strip flexible conduit.

strip conduit is good for use where considerable pulling and bending is required. Its flattened outer surface makes it easier to draw through bored holes, and its inside surface is smooth and even, which allows wires to slide easily during fishing procedures.

On the other hand, double-strip flexible conduit has double armor for protection of the conductors, and may be preferred for this reason. For quick and easy installation, various fittings are used. The fittings can be used with armored cable as well as with flexible conduit. Fittings can be classified as follows:

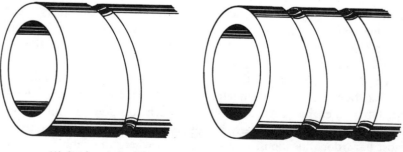

(A) Single-strip bushing.   (B) Double-strip bushing.

Fig. 12. Flexible conduit bushings.

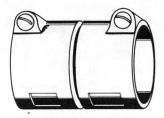

(A) Flexible-to-flexible squeeze-type coupler.

(B) Flexible-to-flexible tangent-screw coupler.

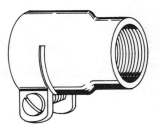

(C) Flexible-to-rigid squeeze-type coupler

(D) Flexible-to-rigid coupler.

(E) Flexible-to-flexible coupler.

Fig. 13. Flexible conduit couplers.

313

(A) Squeeze.

(B) Set-screw.

(C) Slip-in.

(D) Duplex.

(E) 45° angle.

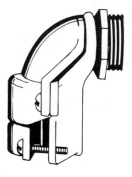

(F) 90° angle.

Fig. 14. Various type connectors.

1. Terminal bushings.
2. Couplings.
3. Connectors.
4. Adapters.

TERMINAL BUSHINGS OR FERRULES

Fig. 12 shows typical brass terminal bushings or ferrules. These bushings are used to protect the conductors from damage due to the raw edge of the conduit. A *coupling* is used to connect two lengths of conduit. Various types of couplings are: Single-strip flexible to single-strip flexible conduit; double-strip flexible to double-strip flexible conduit; single-strip flexible to rigid conduit; double-strip flexible to rigid

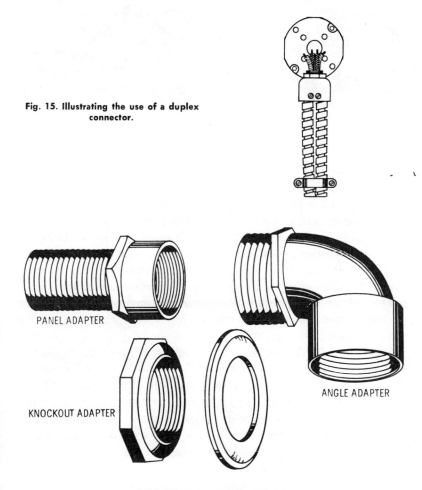

Fig. 15. Illustrating the use of a duplex connector.

PANEL ADAPTER

KNOCKOUT ADAPTER

ANGLE ADAPTER

Fig. 16. Various connector adapters.

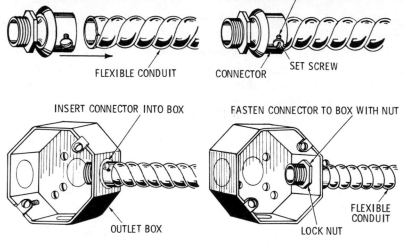

SLIP CONNECTOR OVER CONDUIT

ATTACH CONNECTOR TO CONDUIT WITH

FLEXIBLE CONDUIT

CONNECTOR

SET SCREW

INSERT CONNECTOR INTO BOX

FASTEN CONNECTOR TO BOX WITH NUT

OUTLET BOX

FLEXIBLE CONDUIT

LOCK NUT

Fig. 17. Connecting flexible conduit to outlet box.

conduit. Examples are shown in Fig. 13. Couplings are an essential part of any flexible-conduit installation.

Connectors are fittings used for connecting the conduit to devices such as outlet boxes. Various types are classified as: Squeeze, set-screw, slip-in, duplex, and angle types. Examples are shown in Fig. 14. *Squeeze connectors* give a firm grip entirely around the armor and are supplied with locknuts. *Set-screw connectors* employ a long tangential screw which is forced in the groove of the armor between the armor and the side of the connector. *Slip connectors* are used to save time; no locknut is used and the set-screw is simply tightened after the connector is pushed into the knockout.

Duplex connectors are used in installations where two cables are run into one knockout, as shown in Fig. 15. The use of one duplex connector instead of two separate connectors in two different knockouts saves time and cable, and also doubles the effective number of outlets. Duplex connectors are of tangential set-screw type. *Angle connectors* are used to install conduit at an angle to devices such as outlet boxes. Both 45° and 90° angles are available, as shown in Fig. 14E and F. A few of these connectors will save time and make a better job in cramped locations.

An *adapter,* such as shown in Fig. 16, is a fitting used with rigid conduit. A short section of rigid conduit may use an adapter at one end and a coupling at the other, as shown in Fig. 13D. A summary of the operations in attaching *flexible* conduit to an outlet box is shown in Fig. 17. (Compare with Fig. 15.)

CIRCUITING DETAILS

Circuiting details require careful consideration. Otherwise, after an installation is completed, it may be discovered that the lamp-control facilities are something else than were actually desired. Fig. 18 shows circuit classifications on the basis of current demand. The chief lamp-control arrangements may be listed as follows:

1. A certain number of lamps are to be switched on or off from one location.
2. A pair of lamps are to be individually controlled by a pair of switches.
3. A pair of lamps are to be individually controlled by a single switch. In one arrangement, a main switch must be thrown to turn both lamps off.
4. A pair of lamps are to be individually controlled from one location. In one arrangement, a main switch must be thrown to turn both lamps off.
5. A pair of lamps are to be switched on or off from two locations.
6. Three lamps and three switches are to be wired so that a switch will operate its own light for certain positions of the two other switches.

Fig. 19 shows possible circuit arrangements. Three lamps are controlled from one position in A. Of course, any number of lamps may be used provided the switch and conductors are suitably rated. In B, each lamp is controlled by its own switch; the switches may be in a single unit, or separate units. Many types of switches are in common use; two basic types are shown in Fig. 20. Each lamp in Fig. 19B may be a bank of lamps connected in parallel provided the switches and conductors are suitably rated.

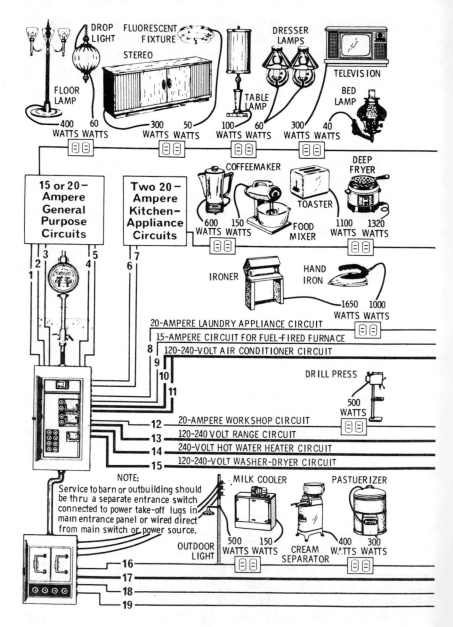

Fig. 18. Classification of circuiting

CEILING LIGHT

ELECTRIC SHAVER

LAMPS

DINING LIGHT

ROOM HEATER

SUNLAMP

TABLE FAN

WORK LIGHT

VACUUM CLEANER

100 WATTS   10 WATTS   275 WATTS   80 WATTS   1600 WATTS   150 WATTS   75 WATTS   80 WATTS   400 WATTS

LARGE GRILL   LARGE ROASTER   STOVE

BLENDER   ROTISSERIE   REFRIGERATOR

1300 WATTS   250 WATTS   1380 WATTS   1400 WATTS   1650 WATTS   250 WATTS

SUMP PUMP   FUEL-FIRED FURNACE   AIR CONDITIONER

300 WATTS   800 WATTS   900 WATTS

WASHER-REG. DRYER

HOT WATER HEATER

POWER SAW   WOOD AND METAL LATHE   ELECTRIC RANGE

570 WATTS   300 WATTS   800 WATTS   2500 WATTS   5200 WATTS

MILKER   HAMMER MILL   CHICK BROODER   ULTRA-VIOLET RAY LAMP

DROP LIGHT

PIG BROODER

WATER PUMP

400 WATTS   1900 WATTS   300 WATTS   1000 WATTS   300 WATTS   125 WATTS   60 WATTS

according to current demand.

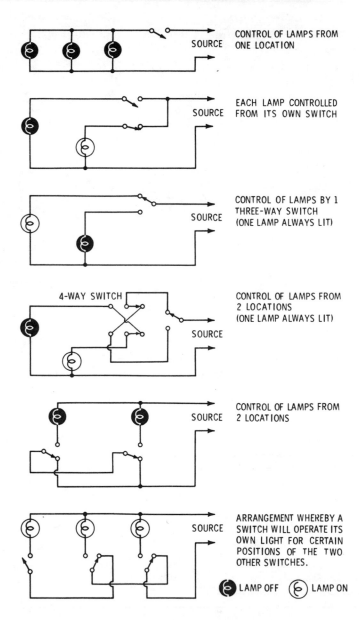

Fig. 19. Possible circuit arrangements using various type switches.

In Fig. 19C a pair of lamps are individually controlled by one 3-way switch. Only one lamp will be on at a time, and both lamps will be dark only if a main switch (not shown) is opened to remove current from the circuit. A 3-way switch is basically a reversing switch. In Fig. 19D, a pair of lamps are individually controlled by a 3-way switch and a 4-way switch at separate locations. Either of the lamps can be turned on from each location, but one of the lamps will always be burning unless a main switch (not shown) is opened.

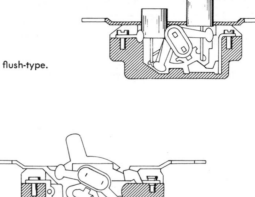

(A) Push-button flush-type.

(B) Toggle flush-type.

Fig. 20. Two basic types of switches.

In Fig. 19E, a pair of lamps are connected in parallel and can be turned on or off by 3-way switches in two separate locations. A larger number of lamps can be connected in parallel if desired provided that the switches and conductors are suitably rated. In Fig. 19F, three lamps and three switches are installed in separate locations. The third light can always be turned on or off by means of the 3-way switch. However, the second light cannot be turned on unless the third light is off. Similarly, the first light cannot be turned on unless both the second and the third lights are turned off. Circuits A, B, and E are in most common use.

Another useful circuit is shown in Fig. 21. This is an arrangement for lamp control from two separate locations. In A, two 3-way switches

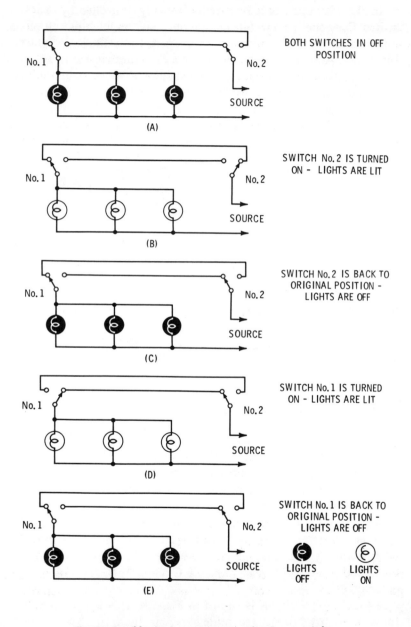

No. 1

No. 2

BOTH SWITCHES IN OFF
POSITION

SOURCE

(A)

No. 1

No. 2

SWITCH No. 2 IS TURNED
ON – LIGHTS ARE LIT

SOURCE

(B)

No. 1

No. 2

SWITCH No. 2 IS BACK TO
ORIGINAL POSITION –
LIGHTS ARE OFF

SOURCE

(C)

No. 1

No. 2

SWITCH No. 1 IS TURNED
ON – LIGHTS ARE LIT

SOURCE

(D)

No. 1

No. 2

SWITCH No. 1 IS BACK TO
ORIGINAL POSITION –
LIGHTS ARE OFF

SOURCE

LIGHTS
OFF

LIGHTS
ON

(E)

Fig. 21. Possible circuit arrangements using 3-way switches.

are used to turn the lamps on or off from either location regardless of the switch setting at the other location. With both switches off, as shown in Fig. 21A, the lamps are dark. Next, if switch No. 2 is turned on, as shown in B, the lamps are turned on. If switch No. 2 is turned off, the lamps are turned off, as shown in Fig. 21C. Then, if switch No. 1 is turned on, as shown in D, the lamps are turned on. If switch No. 1 is turned off, the lamps are turned off.

Note in Fig. 21B that the lamps will be turned off if switch No. 1 is turned off. Similarly, the lamps in D will be turned off if switch No. 2 is turned off. In summary, when either switch is thrown, the lamps will be turned on or off as the case may be. However, the *direction* that a switch must be thrown to turn a lamp on (or off) depends on the setting of the other switch. In other words, the 3-way switch does not have a position marked on or off.

LAMP CONTROL FROM ELECTROLIER SWITCH

The switch shown in Fig. 22 is called a *2-circuit* switch. It is widely used in electroliers. An electrolier is a support for a lamp; for example, a pole lamp, a table lamp, a floor lamp, or a chandelier supported by an electrolier. A 2-circuit switch permits the user to turn two groups of lamps on or off, either in individual groups, or as a whole. For example, a sequence of operation is shown in Fig. 22. In the first position of the switch, both groups of lights are off. In the second position of the switch, the second group of lights is on and the first group is off. In the third position of the switch, both groups of lights are on. In the fourth position of the switch, the first group of lights is on and the second group is off. This switching circuit is not considered as standard, it is only one of several arrangements.

To control three groups of lights, a 3-circuit electrolier switch is used, as shown in Fig. 23. In this arrangement, the groups of lamps are not individually controllable. Instead, each group can be switched on progressively or all groups switched off simultaneously. With the switch in the first position, all lamps are turned off. In the second position of the switch, the second group of lamps is turned on. In the third position of the switch, both the first and second groups of lamps are turned on. In the fourth position of the switch, all three groups of lamps are turned on.

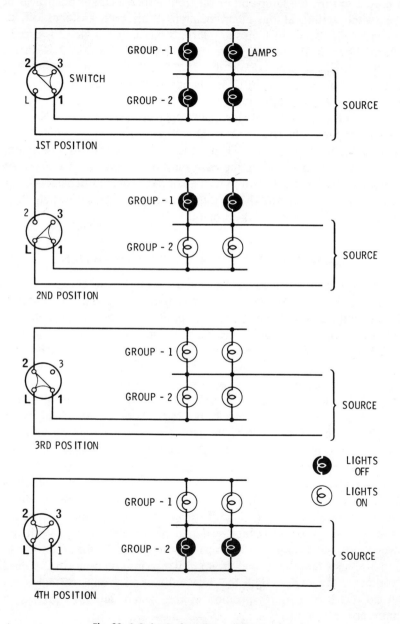

Fig. 22. A 2-circuit electrolier switch arrangement.

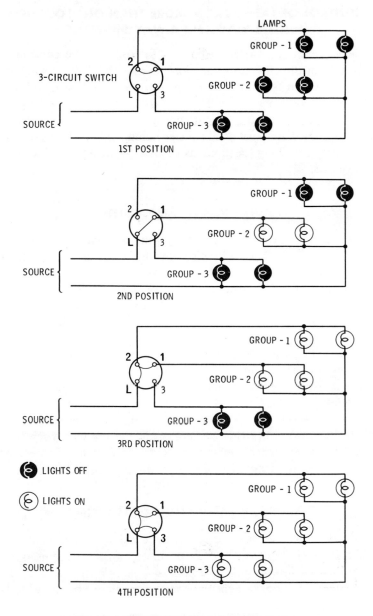

Fig. 23. A 3-circuit electrolier switch arrangement.

325

CONTROL OF LAMPS FROM MORE THAN ONE LOCATION WITH 3-WAY AND 4-WAY SWITCHES

When a lamp (or a parallel bank of lamps) is to be controlled by switches from three or more locations, 3-way and 4-way switches are used as shown in Fig. 24. The lamps can be turned on or off with a switch at any location. Either 3-way or 4-way switches can be used at the ends of the switching circuit; however, 4-way switches must be used at intermediate positions in the switching circuit. It is possible to use as many switching locations as desired by wiring in additional 4-way switches.

CENTRAL POINT LAMP CONTROL

In a central-control lighting system, a master switch is provided, as shown in Fig. 25A. This master switch turns the lights on or off, just as the 3-way and 4-way switches turn the lights on or off. However, when the master switch is on, the other switches cannot turn the lights off. When the master switch is off, the other switches can turn the lights on. Thus, the master switch has complete control of the lights, and the other switches have partial control of the lights.

When the load exceeds the rating of a master switch, a special switch can be used as shown in Fig. 25B. In this arrangement, each switch has partial control of the individual lamps. For example, when both switches are turned off, as in B, all of the lamps are dark. If the special switch is then turned on, all three of the lamps are on, as shown in Fig. 25C. In this position of the special switch, the master switch has no control. However, when the special switch is turned off, the master switch will turn the lamp in its own circuit either on or off.

STAIRWAY LAMP-CONTROL LIGHTING

Three-way switches are widely used in stairway lamp-control circuits, as shown in Fig. 26. Double-pole master switches are used at the ends of the switching circuit. If a person is going upstairs, he operates the switch on one floor to light the lamp on the floor above and to turn out the light on the floor below. In going downstairs, the switching process is reversed. Before installing any switching system, it is im-

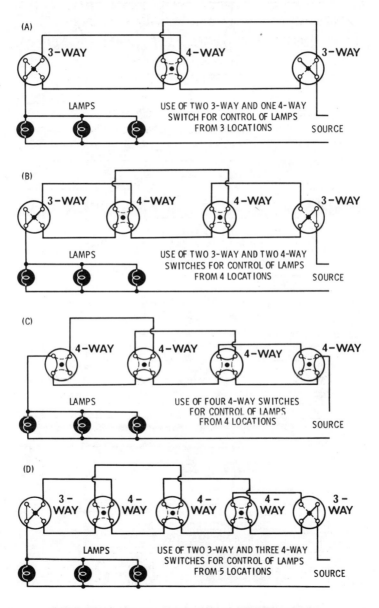

(A) 3-WAY 4-WAY 3-WAY

LAMPS

USE OF TWO 3-WAY AND ONE 4-WAY
SWITCH FOR CONTROL OF LAMPS
FROM 3 LOCATIONS

SOURCE

(B) 3-WAY 4-WAY 4-WAY 3-WAY

LAMPS

USE OF TWO 3-WAY AND TWO 4-WAY
SWITCHES FOR CONTROL OF LAMPS
FROM 4 LOCATIONS

SOURCE

(C) 4-WAY 4-WAY 4-WAY 4-WAY

LAMPS

USE OF FOUR 4-WAY SWITCHES
FOR CONTROL OF LAMPS
FROM 4 LOCATIONS

SOURCE

(D) 3-WAY 4-WAY 4-WAY 4-WAY 3-WAY

LAMPS

USE OF TWO 3-WAY AND THREE 4-WAY
SWITCHES FOR CONTROL OF LAMPS
FROM 5 LOCATIONS

SOURCE

Fig. 24. Switch arrangement using 3-way and 4-way switches.

327

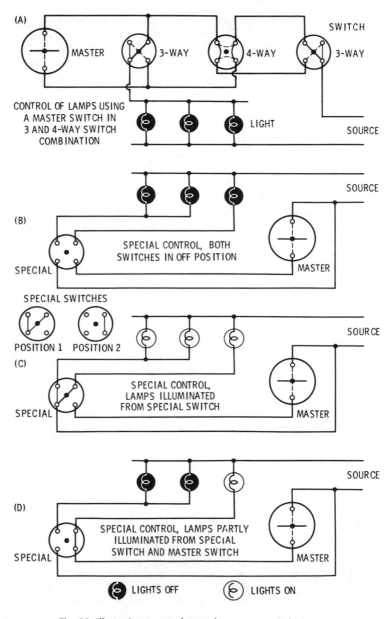

Fig. 25. Illustrating a central-control or master switch system.

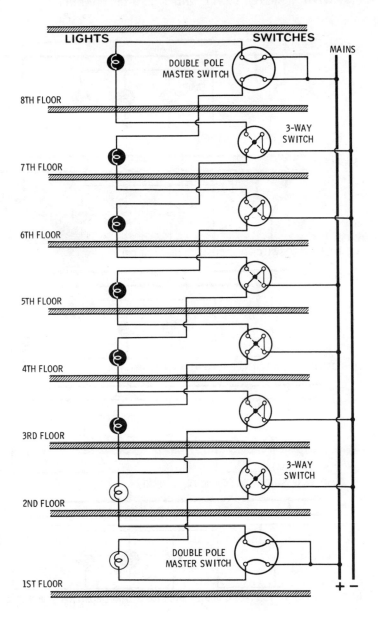

Fig. 26. Possible circuit arrangement using master switches and 3-way switches.

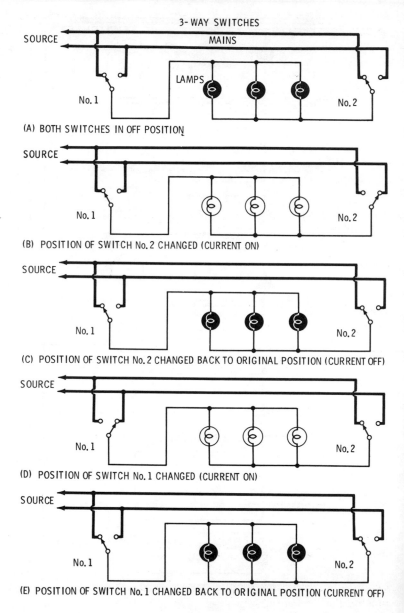

3- WAY SWITCHES

SOURCE

MAINS

LAMPS

No. 1

No. 2

(A) BOTH SWITCHES IN OFF POSITION

SOURCE

No. 1

No. 2

(B) POSITION OF SWITCH No. 2 CHANGED (CURRENT ON)

SOURCE

No. 1

No. 2

(C) POSITION OF SWITCH No. 2 CHANGED BACK TO ORIGINAL POSITION (CURRENT OFF)

SOURCE

No. 1

No. 2

(D) POSITION OF SWITCH No. 1 CHANGED (CURRENT ON)

SOURCE

No. 1

No. 2

(E) POSITION OF SWITCH No. 1 CHANGED BACK TO ORIGINAL POSITION (CURRENT OFF)

Fig. 27. A 3-way switch circuit.

portant to check the applicable electrical codes. As an example of a forbidden switching system, Fig. 27 shows a circuit for lamp control from two locations using the 3-way switches connected in series with a lamp bank.

COLOR CODING OF SWITCH WIRING

As noted previously, black wires are always connected to black wires, white wires are always connected to white wires, and so on. However, when three-way and four-way switches are installed, maintaining the color code requires that white wires from the switches must be painted black at the ends, both at the switches and at the outlet. Fig. 28 shows typical examples of this requirement. Note that the switch terminals marked A and B are light-colored terminals, to which red and white wires must be connected. On the other hand, terminal C is the dark-colored terminal, to which a black wire must be connected. (A light-colored terminal may be a brass terminal in some switches). It is evident from these examples that the color code in the completed installation would be incorrect unless the electrician paints the ends of the white wire black both at the switch and at the light outlet.

SIDE-WIRED AND BACK-WIRED SWITCHES

Most switches have side terminals, as depicted in Fig. 29A, and are called side-wired switches. However, other switches have holes in the back, through which the end of a wire is inserted, and secured by tightening a screw as seen in Fig. 29B. This type of switch is called a back-wired switch. Some switches are designed for either side wiring or back wiring. The advantage of back wiring is that it is easier and faster to merely insert the end of a wire into the switch, than to loop it around a terminal screw. In another design of back-wired switch, a seizing device automatically secures the wire when it is inserted; in other words, there is no screw to be tightened. To release the wire in this type of switch, a metal tab is provided which opens the seizing device when pressed firmly. Fig. 29C shows a switch designed for either back wiring or side wiring. The strip gage indicates how much insula-

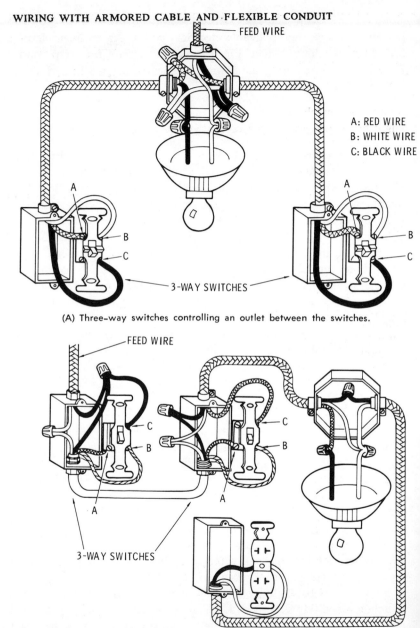

FEED WIRE

A: RED WIRE
B: WHITE WIRE
C: BLACK WIRE

3-WAY SWITCHES

(A) Three-way switches controlling an outlet between the switches.

FEED WIRE

3-WAY SWITCHES

(B) Three-way switches controlling an outlet beyond the switches.

Fig. 28. Color coding

tion should be removed from the end of the wire. Plaster ears are provided to let the switch rest at its correct depth position in case the plaster is thicker than the box setback. Note also that slots are provided for the captive mounting screws. These slots permit the switch to be installed vertically, even though the box might have been mounted to one side or other of perpendicular.

CEILING BOX AND FIXTURE MOUNTING

As depicted in Fig. 30, fixtures are attached to outlet boxes in various ways. For example, an outlet box with a central stud requires only an adapter and a nipple to attach the fixture. Ceiling drop fixtures are usually mounted on a stud, using two nipples and a "hickey" to join them. If the box does not provide a stud, it will have threaded ears to which a strap can be attached for securing the fixture. A strap can also be installed in a box with a stud, using a threaded nipple and locknut. A fluorescent fixture may be mounted on the ceiling with a stud, nipple, and strap. An extension nipple can be used in case the stud happens to be too short for a certain stud and nipple assembly.

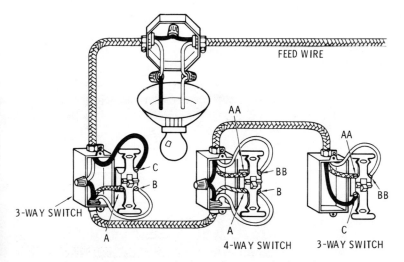

(C) Three- and four-way switches controlling an outlet from three locations.

of switch wiring.

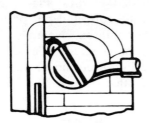

(A) Side-wiring of a switch.

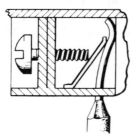

(B) Back-wiring of a switch.

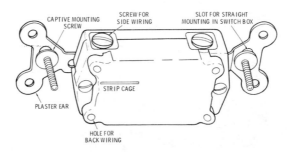

(C) Switch designed for either side-wiring or back-wiring.

Fig. 29. Typical features of switch construction.

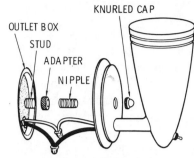

OUTLET BOX
STUD
ADAPTER
NIPPLE
KNURLED CAP

(A) Outlet box with central stud.

OUTLET BOX
SOLDERLESS
CONNECTOR
LOCKNUT
NIPPLE
FIXTURE STUD
HICKEY
CANOPY
COLLAR

(B) Ceiling drop fixture mounting.

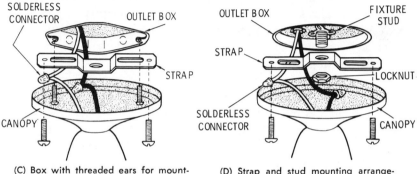

SOLDERLESS
CONNECTOR
OUTLET BOX
STRAP
CANOPY

(C) Box with threaded ears for mounting strap.

OUTLET BOX
STRAP
SOLDERLESS
CONNECTOR
FIXTURE STUD
LOCKNUT
CANOPY

(D) Strap and stud mounting arrangement.

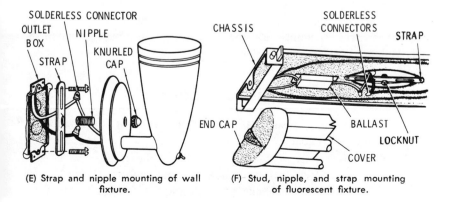

SOLDERLESS CONNECTOR
OUTLET BOX
STRAP
NIPPLE
KNURLED CAP

(E) Strap and nipple mounting of wall fixture.

CHASSIS
SOLDERLESS CONNECTORS
STRAP
END CAP
BALLAST
LOCKNUT
COVER

(F) Stud, nipple, and strap mounting of fluorescent fixture.

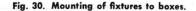

Fig. 30. Mounting of fixtures to boxes.

SUMMARY

Armored cable does not have the advantages of the conduit installation where old wires can be removed and new wires inserted, but armored cable is flexible and can be fished between floors and between partition walls. It is also less expensive than rigid conduit or flexible steel conduit. There are many buildings where the installation of armored cable is not permitted.

Flexible conduit consists of a continuous flexible steel tube which is made from convex and concave metal strips in a spiral. Flexible conduit is also restricted in various installations where oil, gasoline, or other chemicals are used. Although flexible tubing is easy to work with, it is seldom desirable to install an entire wiring job with it. It is usually combined in short runs with rigid conduit.

When wiring a house, the light control system must be carefully considered. Otherwise, after an installation is completed, it may be discovered that the lamp-control facilities are something else than were actually desired. When lighting is to be controlled by switches from three or more locations, 3-way and 4-way switches are used. The lighting can be turned on or off with a switch at any location. Either 3-way or 4-way switches can be used at the end of the switching circuit; however, 4-way switches must be used at intermediate positions in the switching circuits.

In a central-control lighting system, a master switch is provided to control the lights. However, when the master switch is on, the other switches cannot turn the lights off—the master switch has complete control.

TEST QUESTIONS

1. What colors of insulation are used in a two-wire armored cable? In a 3-wire cable?
2. Where would lead-covered armored cable be used?
3. Why is a bonding strip used in an armored cable?
4. How is armored cable cut for installation?
5. What is the purpose of an armored-cable bushing?
6. When is a knockout closer used?

7. Why is a chain used for fishing armored cable?
8. Explain the shortest bend that is allowed in armored cable by the National Electrical Code.
9. Name several applications in which armored cable is forbidden.
10. How is armored cable passed around braces? Across doorways?
11. Describe the difference between armored cable and flexible conduit.
12. Name the two basic types of flexible conduit.
13. What types of couplings are used with flexible conduit?
14. Describe two basic types of connectors.
15. Explain how flexible conduit can be attached to an outlet box.
16. What is a 3-way switch?
17. Discuss a simple application for a 3-way switch.
18. What is a 4-way switch?
19. How would a 4-way switch be used in a circuit with a 3-way switch?
20. Explain how switches are used to control a lamp or a group of lamps from two separate locations.
21. What is an electrolier?
22. Describe a 2-circuit switch.
23. How is a 3-circuit electrolier switch wired to control three groups of lamps?
24. How would a circuit be wired to control a lamp bank from five separate locations?
25. Explain what is meant by central point lamp control.

CHAPTER 12

Wiring Requirements for the Home

In planning an electrical system, the first consideration is *safety,* and the next consideration is *function.* The functioning of a system should permit full and convenient use of both present and future electrical equipment. An effective and efficient home wiring system depends on:

1. Sufficient circuits of adequately large wire to supply the various loads without uneconomical voltage drop.

2. A satisfactory number of outlets to permit convenient use of electrical equipment.

3. High-quality materials and good workmanship.

PLANNING OF EQUIPMENT

The size of a house is directly related to the extent of the electrical installation. Appliances are usually classified as fixed or portable. The owner must determine the type and number of lighting units and appliances to be used. Typical fixed appliances are water heaters, washers, ironers, dryers, ranges, refrigerators, air conditioners, dishwashers, garbage disposers, etc. If heating and air-conditioning units are to be installed, provision must be made in the wiring plan for blowers, fur-

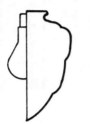

(A) Opal-glass diffusing globe.

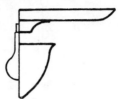

(B) Planetlite luminaire.

(C) Keldon semi-indirect luminaire.

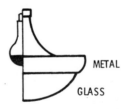

METAL

GLASS

(D) Indirect luminaire.

(E) Silvered-glass reflector.

(F) Symmetric dome reflector.

Fig. 1. Various shapes of light fixtures.

nace motors, etc. Typical portable appliances are electric clocks, coffee makers, fans, irons, heating blankets, mixers, radios, television receivers, roasters, sewing machines, shavers, sun lamps, toasters, and vacuum cleaners.

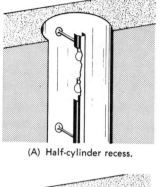

(A) Half-cylinder recess.

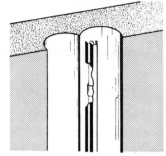

(B) Double recess.

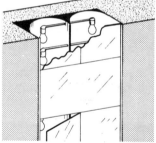

(C) Flush panel.

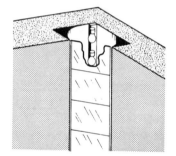

(D) Corner panel.

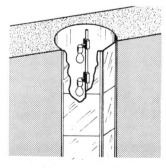

(E) Projecting elements.

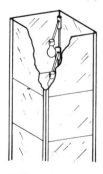

(F) Square column.

(G) Cylinder.

Fig. 2. Built-in lighting.

Lighting facilities must meet three basic requirements:

1. There must be sufficient light for all activities from the viewpoint of comfort and avoidance of eye fatigue.
2. Light must be provided so as to avoid objectionable shadows and controlled to eliminate glare.
3. Fixtures must be adapted to their purpose and appropriate to their surroundings. Fig. 1 shows outlines of typical luminaires. Note that the type of luminaire will partially determine how many will be required for a given area.

Built-in lighting has been briefly noted previously. It comprises luminous panels, columns, or similar structures on existing walls, or it may be part of the house architecturally. Fig. 2 shows some examples of built-in lighting. Lamps range from general-service, tubular, and

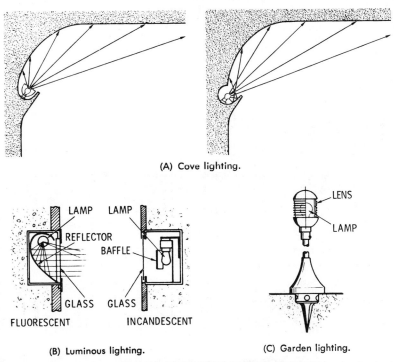

(A) Cove lighting.

(B) Luminous lighting.

(C) Garden lighting.

Fig. 3. Examples of light projection.

Lumiline types to specialized tungsten lamps. Lumiline is a thin tubular type of fluorescent lamp that starts instantly, as explained subsequently. Cove lighting is shown in Fig. 3A. Entrance lighting should clearly illuminate the path, steps, house number, and bell switch; see Fig. 3B and C. Weatherproof lanterns or built-in lighting units are often used.

If a house is located some distance from the sidewalk, an illuminated number may be installed at the driveway entrance. Small floodlights are sometimes placed in shrubbery to illuminate the front of the house. Entrance fixtures should be controlled by switches installed just outside the front door; switches should be on the lock side—not on the hinge side. Fig. 4 shows typical switches. Night lights make it easy to find a switch in the dark. Pilot lights serve as a useful reminder that a switch is turned on. Weatherproof convenience outlets should be installed outside of the house for electrical garden equipment and seasonal lighting. A vestibule between the hall and entrance is usually small, and can be illuminated by a ceiling fixture with a switch adjacent to the entrance switch.

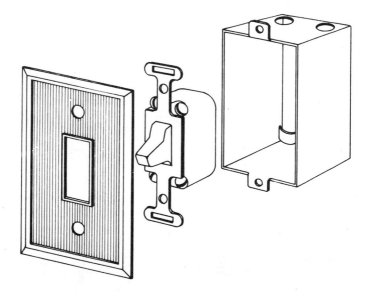

Fig. 4. Wall switch box, switch, and face plate.

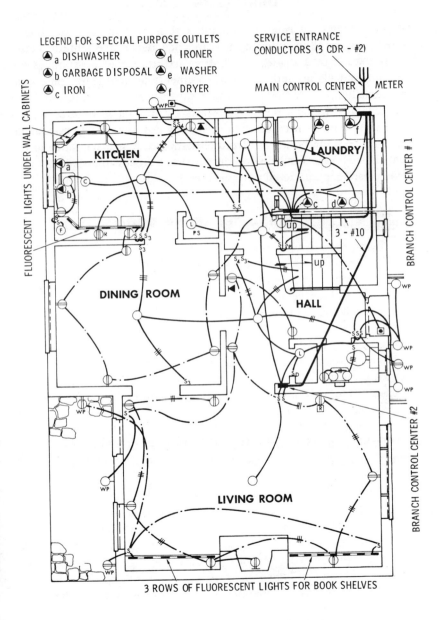

LEGEND FOR SPECIAL PURPOSE OUTLETS
- ⚤ₐ DISHWASHER
- ⚤_d IRONER
- ⚤_b GARBAGE DISPOSAL
- ⚤_e WASHER
- ⚤_c IRON
- ⚤_f DRYER

SERVICE ENTRANCE CONDUCTORS (3 CDR - #2)

MAIN CONTROL CENTER

METER

FLUORESCENT LIGHTS UNDER WALL CABINETS

KITCHEN

LAUNDRY

BRANCH CONTROL CENTER # 1

DINING ROOM

HALL

3 - #10

up

up

LIVING ROOM

BRANCH CONTROL CENTER #2

3 ROWS OF FLUORESCENT LIGHTS FOR BOOK SHELVES

Fig. 5. Wiring plan for first floor of a two-story house.

343

LIGHTING THE HALLS

Fig. 5 shows a wiring plan for the first story of a two-story home. A hall may be illuminated by a suspended decorative lantern, close-fitting fixture, or a shaded candle-type fixture. Other possibilities are luminous panels around a mirror, torchieres, a pole lamp, or a console table lamp. Note that light from a downstairs hall fixture should be supplemented by light from an upstairs fixture. These fixtures are usually controlled by separate switches, both upstairs and down, so that they can be turned on or off from either floor. It is essential to install convenience outlets in halls for electrical appliances; these outlets may also be used for additional portable lamps.

LIGHTING THE LIVING ROOM

A living room usually has the brightest illumination level, with the possible exception of a library or studio. Light sources for a living room are generally divided between fixed and portable units. Fixed units may include ceiling fixtures, wall brackets, cove lights, recess lights, or column lights. Portable units may include floor lamps, table lamps, small high-intensity lamps, or semiportable units such as pole lamps.

Although fewer ceiling fixtures are used in many installations, they are not a thing of the past. Pendant fixtures such as chandeliers may be used in the more elaborate installations. Note that rooms with low ceilings require close-fitting and comparatively closely-spaced ceiling fixtures, or built-in fixtures. Reasonably uniform general illumination is desirable in a living room. High-level local illumination for reading, sewing, or for piano or organ playing is obtained by various types of supplementary lighting units.

In planning the system, keep in mind that wall brackets and totally-indirect wall urns have often-needed decorative value as well as contributing to greater uniformity of general room illumination. Wall brackets must be located so that they do not interfere with tapestries or paintings. Three-way lamps (also called 3-way lights) are often used for supplementary illumination at sofas. A 3-way light bulb has two filaments which may be operated together or separately. They generally provide 100-, 200-, or 300-watt levels from a switch with three

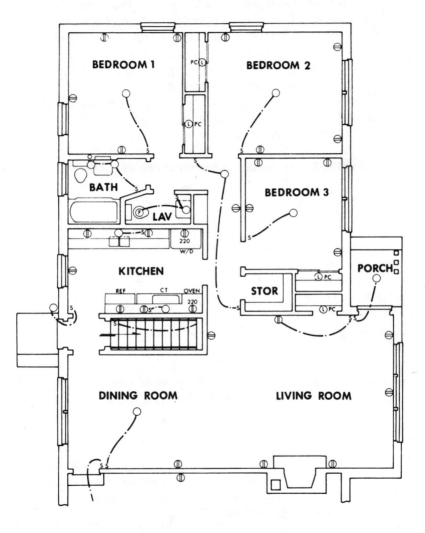

Fig. 6. Floor plan of a house showing wiring diagram.

345

"on" positions. The lamp is used in a mogul 3-way socket which provides one contact at the side of the base plus a button contact and a ring contact at the base.

Note that every wall space in a living room that is large enough for a table, easy chair, or television receiver, should be provided with a duplex convenience outlet. In any case, a duplex outlet should be installed at every 12-foot interval, or less, along a living-room wall. Also, each mantel should have at least one flush-mounted outlet on its top for an electric clock or decorative light.

A common fault in living-room lighting plans is lack of *functional* switching facilities. Switch location and switching circuits should be planned for greatest convenience in everyday use. Good planning requires careful attention to the architectural layout. For example, the plan shown in Fig. 6 requires different switching facilities than the plan shown in Fig. 5. Chandeliers, ceiling fixtures, or wall brackets are usually controlled by a switch at the main entrance of the living room and at each subsidiary entrance. Note that an outlet for an electric clock should not be connected to a switch.

Whenever possible, switching facilities should be discussed in detail with the home-owner and double-checked. To avoid overlooking various requirements, the planners should go over the rooms in the house step-by-step at twilight instead of broad daylight. Then, by "sizing up" the plan from each entrance, and from each seating location, functional requirements can be recognized to best advantage. All too often, electricians and home-owners plan switching installations in haste.

LIGHTING THE DINING ROOM

Dining tables are often illuminated by pendant fixture, such as shown in Fig. 7. Wall brackets will provide pleasant background illumination, if properly shaded. Electricians work with lower ceiling lights than formerly, and with combined living-room and dining areas, as in Fig. 6. In turn, more careful planning is required to obtain good lighting proportions and balance. Colored lighting, such as amber or moonlight tones, may be used for general lighting, but this must be supplemented by white localized lighting at any table that may be used for study or close work.

Fig. 7. A pendant fixture.

Many interesting effects can be obtained by installations of white or colored lumiline or larger fluorescent lamps installed in plaster, metal, or glass coves around a dining room, or at opposite ends of a living room. Aside from provision for illumination, convenience outlets must be installed for electrical appliances on the table or at the buffet. Lack of foresight can result in the necessity for running extension cords over floors or under rugs where they are a continuing nuisance. Switching facilities for a dining room should follow a functional plan, as explained previously for living rooms.

LIGHTING THE KITCHEN

Kitchen lighting should be both cheery and utilitarian. A central ceiling fixture is almost always required. This is usually a single diffusing enclosed unit. Modern-styled kitchens may have totally indirect fixtures. If the walls and ceiling have a light-colored surface, the problem of good kitchen lighting is more easily solved. The general rule is that plenty of light should be provided at any point that may be occupied. Therefore, local lighting is necessary over the range, sink, and associated work areas.

In most installations, the first requirement is a pendant fixture over the sink and range. A glass shade is essential to protect the cook from glare. There is a marked trend in modern kitchens to employ recessed units covered by diffusing glass panels. *Soffit lighting* can be defined as a tailored installation, which may be recessed into furring above a kitchen sink. Fig. 8 shows how a ventilator can also be advantageous-

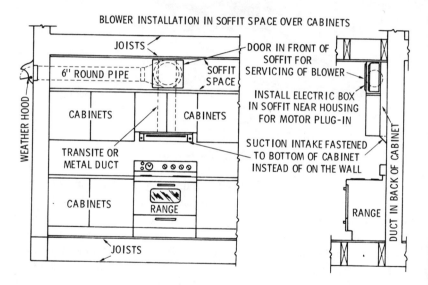

Fig. 8. Soffit with blower mounted in place. Light mounted beside the blower.

ly mounted in soffit space over kitchen cabinets. Lights can often be mounted advantageously beside the blower.

In the foregoing arrangement, fluorescent lamps may be mounted in curved metallic reflectors, with the diffusing element provided by frosted glass covering the *lamp installation* beside the blower. Either lumiline or ordinary fluorescent lamps are good choices for installation over the range, sink, or under the kitchen cabinets to illuminate the worktable or tables. In a butler's pantry, a small close-fitting ceiling fixture is adequate to light shelves or cupboards. In case a breakfast alcove is provided, a pendant-shaped fixture or indirect lighting unit will usually meet illumination requirements.

Obviously, an ample number of convenience outlets must be provided in a kitchen. Toasters, percolators, mixers, and roasters are always used at table height, and their outlets should be installed 42 inches above the floor. On the other hand, electric clocks and ventilating fans will usually be mounted higher, and their outlets will be installed from six to eight feet above floor level in most kitchens. Dishwashers and refrigerators are generally connected to functionally-located outlets. The range is connected to a specialized line as illustrated in Fig. 9.

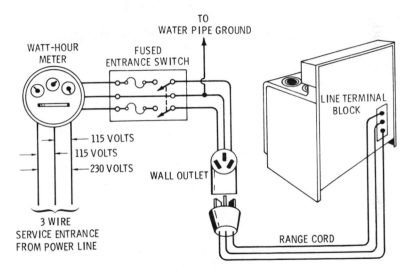

Fig. 9. Typical installation of electric-range wiring.

As explained previously, a 3-wire line is installed from the service entrance as shown in Fig. 10. The neutral wire is grounded. Where substantial power is demanded, 230-volt operation provides greater economy than 115-volt operation, due to copper costs versus line drop. In other words, voltage drop is *directly* proportional to current, whereas power is proportional to *current squared*. Therefore, substantially greater efficiency can be obtained by supplying power at double voltage and half current. As also noted previously, the electrician must

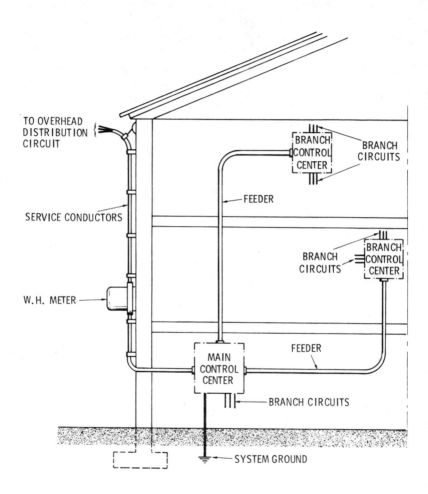

Fig. 10. Pictorial diagram of 3-wire line installation through the service entrance to various locations in the house.

| 60 WATTS | THREE 40 WATTS | 60 WATTS | THREE 60 WATTS | 100 WATTS |

(A) Living room.

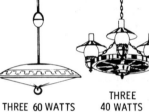

| FIVE 20 WATTS | 100 WATTS | THREE 60 WATTS | THREE 40 WATTS | 100 WATTS |

(B) Dining room.

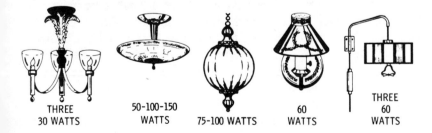

| THREE 30 WATTS | 50-100-150 WATTS | 75-100 WATTS | 60 WATTS | THREE 60 WATTS |

(C) Bedroom.

Fig. 11. Lamp styles used in various rooms by interior decorators.

351

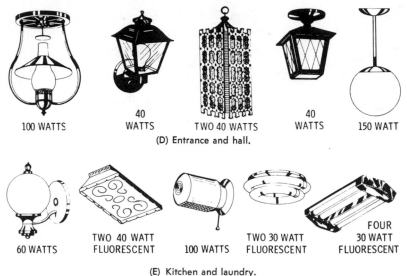

100 WATTS

40 WATTS

TWO 40 WATTS

40 WATTS

150 WATT

(D) Entrance and hall.

60 WATTS

TWO 40 WATT FLUORESCENT

100 WATTS

TWO 30 WATT FLUORESCENT

FOUR 30 WATT FLUORESCENT

(E) Kitchen and laundry.

Fig. 11. Lamp styles used in various rooms by interior decorators (contd).

connect a heavy "jumper" around the water meter if the ground is to pass inspection.

LIGHTING THE BEDROOM

Bedroom illumination tends to be skimped in many installations. For example, low-wattage night lights add considerably to functional utility. A night light may consist merely of a small plug-in electro-luminescent panel in an economy-type installation. Or, a neon glow lamp may be installed near the bathroom or light switch. A bedroom should have a general lighting unit such as a ceiling fixture, supple-mented by local units for reading, or possibly sewing. Most types of boudoir units are inadequate except for decorative lighting.

Wall brackets are not always indicated for bedroom illumination because they may tend to limit furniture arrangements. Ceiling fixtures may be semidirect or totally indirect types. All bulbs must be shaded in a candle-type fixture. Typical lamp styles suggested by interior dec-orators for various rooms are shown in Fig. 11. Note that there are three places in a bedroom that need special lighting; the vanity dresser, the bed, and the boudoir chair or chaise lounge. Boudoir lamps on the vanity should be tall, with light-colored shades to provide adequate lighting on both sides of the face. For reading in bed, a wall, table, or floor lamp can be used if designed so that glaring light does not occur at eye level.

A lamp with a white diffusing bowl under the shade is a good choice in most installations. Also, a floor lamp that provides a wide spread of diffusion light downward beside the boudoir chair or chaise lounge will be suitable for mending, knitting, or reading. A small lamp placed under the bed in the baseboard will provide floor illumination. In turn, convenience outlets for the lamps must be installed in each wall. Closet illumination should not be overlooked. Each closet over two feet in depth should have a light installed inside on the wall or in the ceiling just above the door so that light is directed back on the clothes and shelves.

LIGHTING THE BATHROOM

Bathroom lighting should be planned so that the mirror can be used to advantage for shaving. Two brackets installed about 5½ feet from the floor level may be adequate. The advantage of two units is in minimizing shadows and providing uniform illumination from each side. Bulbs are usually shaded with diffusing glass. In the more elaborate installations, the mirror is framed along the top and the sides by built-in panels with lamps concealed behind frosted glass.

Most bathrooms require a ceiling fixture. A globe-type fixture is typical. Switch facilities should be provided just inside the door, with a separate switch for the mirror light. Note that showers and tubs may need lighting in some cases. These lights are waterproof types and are usually mounted flush with the ceiling. A night light installed in the baseboard provides practical utility, just as in a bedroom. A convenience outlet installed at the right-hand side of the mirror is necessary for use of electric razors, curling irons, or sun lamps.

LIGHTING THE ATTIC

An attic space usually requires a ceiling light and a convenience outlet. A standard dome reflector may be used, controlled by switches

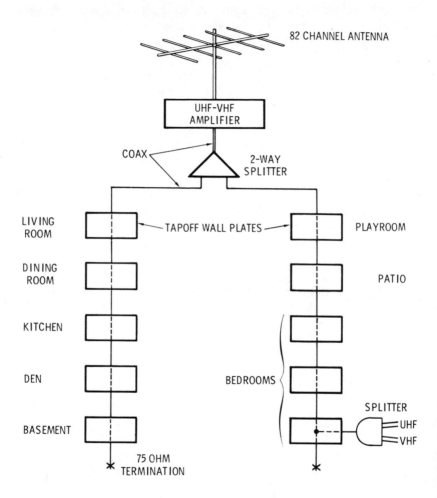

Fig. 12. Small master antenna (MATV) system.

at the foot and at the head of the attic stairs. Additional wiring is required in case an attic fan or blower is installed. Lead-ins for television and radio receivers may be routed through an attic. Fig. 12 shows a wiring plan for a small master-antenna television system. This type of wiring is specialized and may be subcontracted by an electrician in case he does not wish to do this kind of work.

LIGHTING THE BASEMENT

Basement areas may be typically divided into stairway, laundry, workshop, furnace room, and recreation room. A glareless light installed at the foot of the stairs and controlled by 3-way switches from the head of the stairs and from the basement, is usually required. A pilot light in the switch at the head of the stairs is useful as an indicator to show if the light has been left on.

A laundry room needs one or more ceiling lights, depending on the area to be illuminated. White diffusing globe luminaires or silvered bowl lamps are often installed. Local light is almost always required in the laundry and at various appliances. The general rule to be observed is provision of ample light at each work area. Convenience outlets are required at suitable locations for portable appliances.

Workbenches must be well illuminated and shadows avoided. This is particularly important in the operation of power tools. Diffusers installed over a workbench will direct light downward for close mechanical work. Local lighting units are also desirable in many cases. An ample number of well-located duplex outlets should be provided throughout the work area.

Furnace rooms do not require extensive illumination, and a ceiling reflector installed at the furnace is often adequate. The light should clearly illuminate all controls and indicating scales. It is good practice to install at least one convenience outlet for use in maintenance. If a number of units are installed in a furnace room, it may be necessary to supplement a ceiling reflector with one or more local units. The rule to be observed is that of good utility under both normal and maintenance requirements.

Recreation rooms are often associated with the basement area. Low ceilings may have to be contended with, and recessed or close-fitting fixtures may be most suitable. Portable lamps are frequently necessary for card tables, etc. Bars require special lighting installations with more or less decorative effects. An ample number of convenience outlets should be provided for appliances, electric toys, and so on.

LIGHTING THE GARAGE

Garages may be associated with basement or general living areas. In each case, the requirements must be carefully "sized up" by the electrician. Good functional illumination is the prime consideration. Unfortunately, garage lighting is often skimped, which not only imposes a hazard at night, but also complicates any automobile maintenance work which might be required. Good lighting is essential at the rear door from a safety standpoint. This may be a simple bracket or porch-ceiling light. An exterior light is often a decorative waterproof lantern, or a reflector.

Lights should be controlled by switches installed inside the entrance to the garage and inside the most convenient door to the house. Modern garages often have radio-controlled automatic door openers. (See Fig. 13.) Although the wiring of door openers is a specialized job, it

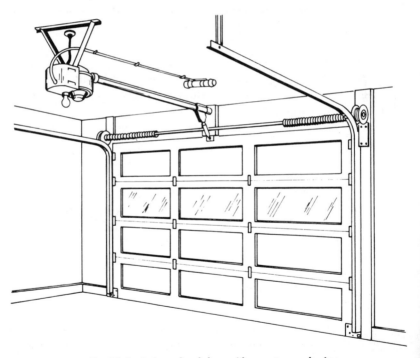

Fig. 13. Typical overhead door with operator mechanism.

is not unduly difficult. Detailed instructions for installation will be provided with the equipment. Convenience outlets should be provided in a garage for use of lamps with extension cords and electrical tools.

MASTER SWITCHES

Master switches are installed chiefly in bedrooms. A master switch makes it possible to turn on lights located at various points inside or outside the house independently of local switches. When the master switch is in its "off" position, the circuits can then be controlled by the local switches. This basic arrangement was noted in the previous chapter. If two banks of lamps in one circuit are to be controlled by a master switch, the lamps and switches are wired as shown in Fig. 14. Or, if two circuits with two banks of lamps in each circuit are used, the connections shown in Fig. 15 are required.

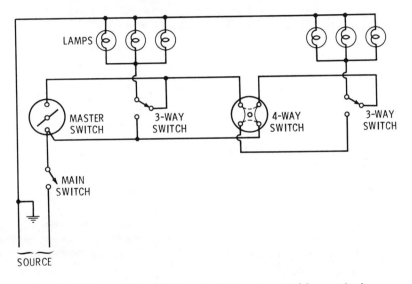

Fig. 14. Typical wiring diagram showing master control for one circuit.

In more elaborate installations, the control facilities in the bedroom consist of switches which can turn various house or outside lamps on or off independently of local switches. Moreover, switching relays and indicator lights are wired into the system so that any lamp, or lamp bank that is "on" causes a particular indicator lamp to glow in the bedroom. In turn, when the homeowner retires for the night, he is reminded of any lights that have been left turned on, and he can turn the lights off from the master switch installation. The wiring is com-

357

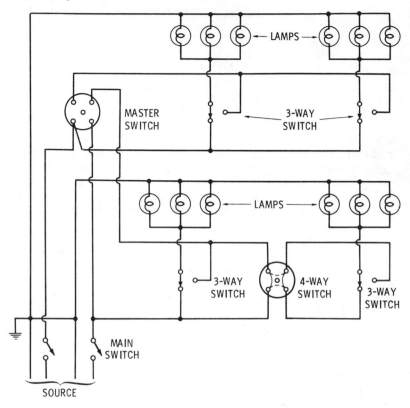

Fig. 15. Typical wiring diagram showing master control for two circuits.

paratively complex, but not unduly difficult. Complete installation instructions are included with the equipment.

In planning a system, wire sizes and switch ratings must be sufficient to meet code regulations. It is important to note that a system might not pass inspection, although wire sizes, outlet, and switch ratings are ample, simply because the components are not *matched*. In other words, suppose that No. 14 wire is sufficient to meet code regulations. If we install No. 12 wire, the system will not pass inspection unless the current rating of the switches and outlets *match* the current-carrying capacity of the wire. It makes no difference from the inspector's point of view that the switches and outlets meet the load requirements —oversize wires must always be connected to matching devices.

358

LAMP AUXILIARIES

Modern fluorescent lamps are available in four types:

1. Hot cathode, preheat starting type.
2. Hot-cathode, instant starting.
3. Cold-cathode.
4. RF lamp.

Fig. 16 shows the circuits that are typical for each type of lamp. When a manual starting switch is used, the preheating period depends on the time that the operator keeps the switch depressed. However, glow-type starters such as shown in Fig. 17 are automatic; they are preferable because the preheating period is closely controlled, with the result that the lamp is usually longer-lived.

A thermal-type starter employs heat from a resistance, instead of neon gas as in a glow-type starter. When the lamp is turned on, the bimetal strip touches a set of carbon contacts. In turn, current flows through the lamp cathodes for preheating. The resistance of the carbon also produces heat which soon causes the bimetal strip to move away from the contacts, thus striking an arc inside the lamp. Thereafter, the bimetal strip is kept away from the contacts by current flowing through the resistor which is in parallel with the lamp. In case the current should be momentarily interrupted, the voltage across the resistor increases and the additional heat causes the bimetal strip to bend until it touches the restart contact. In turn, the cathodes are preheated again, but the resistor is now short-circuited and the bimetal strip soon moves away from the restart contacts and the arc is struck inside the lamp.

In instant-starting lamps, hot cathodes may be used, and the arc is struck immediately by a high kickback voltage from the ballast reactor. The cathodes are specially manufactured to withstand the effects of starting voltages between 450 and 750 volts. Instant-starting lamps also use cold cathodes. This type of lamp has the advantage that its life is not shortened by brief operating periods with frequent operation of the switch. An RF lamp is similar to a low-pressure mercury-arc lamp, except that tubing is used and is coated with a phosphor which converts ultraviolet light into visible light. RF lamps are used to a greater extent in industry, rather than in home lighting systems.

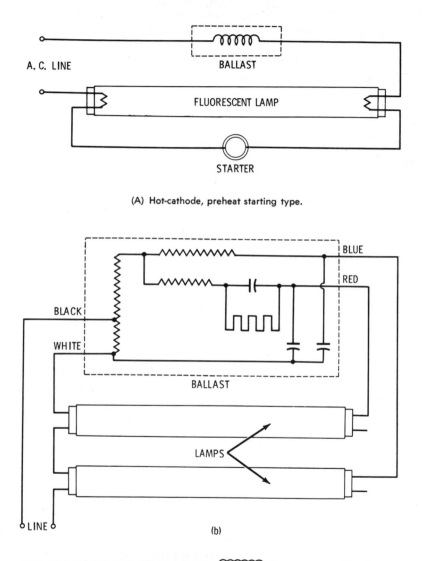

(A) Hot-cathode, preheat starting type.

LAMPS

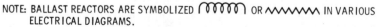

(b)

NOTE: BALLAST REACTORS ARE SYMBOLIZED ⟨⟨⟨⟨⟨⟩⟩⟩⟩⟩ OR ∧∧∧∧∧∧ IN VARIOUS ELECTRICAL DIAGRAMS.

(B) Hot-cathode, instant starting.

Fig. 16. Four types

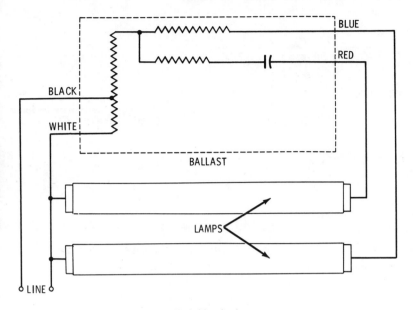

(C) Cold-cathode.

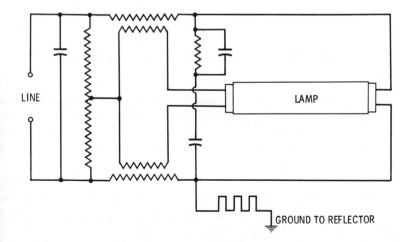

(D) RF lamp.

of fluorescent lamps.

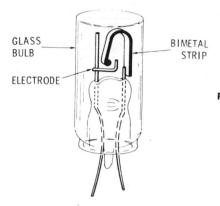

GLASS
BULB

ELECTRODE

BIMETAL
STRIP

Fig. 17. Showing principal parts of typical glow type starter.

Ballast reactors are usually installed in the raceway behind the lamp. A ballast may be used alone or in combination with a step-up transformer to obtain increased starting voltage. Ballasts are also used in combination with capacitors for power-factor correction, so that the lamp circuit draws current in-phase with the applied voltage. Power-factor capacitors are seen in Fig. 16. All of these lamp auxiliaries are usually provided as a single enclosed unit which is mounted in the raceway.

GROUND CIRCUIT FAULT INTERRUPTERS

Fuses do not blow and circuit breakers do not trip until the current demand exceeds the ampacity rating of the fuse or breaker. In turn, if there is electrical leakage to the frame of an appliance such as a lawnmower, for example, a person can be fatally shocked by a very small fraction of one ampere. If three-prong plugs on three-wire cords (cable with a grounded conductor) are used, the chances of receiving a shock is quite small. However, ground connections in wiring systems can become defective. As an illustration, the grounding screw depicted in Fig. 18 might become loose, or the grounding wire could be broken. In addition, many existing wiring systems do not have grounding conductors, and no protection is provided against electrical leakage in appliances. For this reason, the National Electrical Code now requires the installation of ground fault circuit interrupters in

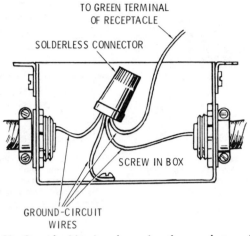

TO GREEN TERMINAL
OF RECEPTACLE

SOLDERLESS CONNECTOR

SCREW IN BOX

GROUND-CIRCUIT
WIRES

Fig. 18. Ground wiring in a box using three-conductor cable.

various locations. A GFCI is a special outlet device that includes a relay which trips on 5 milliamperes of leakage current. In turn, the user of an appliance is protected against defects that make the frame of the appliance "hot". Although 5 milliamperes will shock the user, the relay trips in 0.025 second, so that the shock is felt only for an instant.

FISHING WIRES IN OLD WORK

Old work (additions or modernization) usually requires fishing of wires through walls and across ceilings, as shown in Fig. 19. This procedure can be very difficult unless holes are drilled correctly and the proper steps are followed. Non-metallic sheathed cable or dual-purpose plastic cable must be used. When floor boards or wall boards are replaced, it is very important to avoid driving nails where they could damage the cable. Fishing wires generally requires two people to work the fish wires and catch the hooks together.

SUMMARY

Safety and function are the first considerations in planning an electrical system. An effective and efficient home wiring system

363

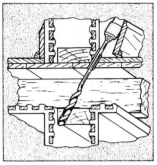

If you can get into attic or upper room, remove the upstairs baseboard. Drill diagonal hole downward as shown.

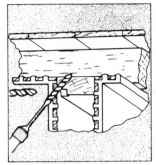

Drill diagonal hole upward from opposite room. Then drill horizontally till holes meet. This procedure requires patching plaster.

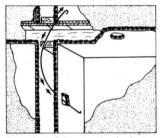

Push 12-foot fish wire, hooked at two ends, through hole on 2nd floor. Pull one end out at outlet on 1st floor.

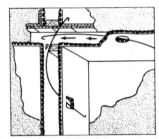

Next, push 20-25-foot fish wire, hooked at both ends thru ceiling outlet (arrows). Work the fish wire until you touch the first wire.

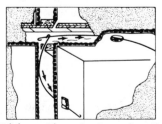

Withdraw either wire (arrows) until it hooks the other wire; then withdraw second wire until both hooks hook together.

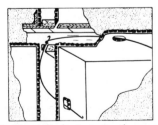

Finally, pull shorter wire thru switch outlet. When hook from long wire appears, attach cable and pull thru wall and ceiling.

Fig. 19. Steps in fishing a cable for an old work installation.

depends on sufficient circuits, satisfactory number of outlets, and high-quality material and workmanship.

Lighting facilities must meet three basic requirements—sufficient lights for all activities, fixtures must be adapted to their purpose and appropriate to the surroundings, and the avoidance of objectionable shadows and glare.

Lighting in the living room generally has the brightest illumination level, and is divided between fixed and portable fixtures. In more elaborate homes, ceiling fixtures or chandeliers are used, but in smaller homes where lower ceilings are generally the design, wall fixtures, table lamps, and pole lighting is used. Many times totally indirect wall lighting is used, giving a uniform room illumination. In this type of lighting, wall brackets must be located so that they do not interfere with tapestries or paintings.

In dining rooms, more careful planning is required to obtain good lighting proportions and balance. Colored lighting, such as amber or moonlight tones, may be used for general lighting, but must be supplemented by white localized lighting at tables which are used for study or close work. Convenience outlets must be installed for electrical appliances used at the table or at the buffet. Lack of foresight can result in the necessity for running extension cords over floors or under rugs, which can be a general nuisance.

Kitchen lighting should be both cheery and utilitarian. A central ceiling light is almost always required, which is usually a single diffusing enclosed unit. There is a trend in modern kitchens to employ recessed units covered by diffusing glass panels which also serves as the kitchen ceiling.

Bedroom and bathroom fixtures should provide adequate lighting at dressing mirrors. There are three places in the bedroom that needs special lighting; the vanity dresser, the bed, and the boudoir chair or chaise lounge. Bathroom lighting should be planned so that the mirror can be used to advantage for shaving. The shower and tub may need lighting in some cases.

The basement may be divided into four or five areas, such as laundry, workshop, furnace room, and recreation room. The laundry and furnace room may need one or more ceiling fixtures depending on the area to be illuminated. White diffusing globes or silvered bowl lamps are often used. Low ceilings may exist in the recreation room, and recessed or close-fitting fixtures may be most suitable.

TEST QUESTIONS

1. What are the three chief requirements for a good home-lighting system?
2. Name several types of luminaires that are suitable for home-lighting installations.
3. How should entrance lighting be planned? Where are switches installed?
4. Discuss the requirements for good hall lighting.
5. Explain how an electrician plans a living-room installation when the ceiling is low.
6. What are the requirements for installation of convenience outlets?
7. Why should colored lighting in a dining room be supplemented by white localized lighting?
8. Describe the requirements for good kitchen lighting.
9. What does an electrician mean by soffit lighting?
10. Why does an electric range require separate wiring?
11. Explain the requirements for night lights in bedrooms and bathrooms.
12. Name several types of lamps that are used in modern bedroom and bathroom installations.
13. What are the general requirements for attic lighting?
14. Why may specialized wiring be installed in an attic?
15. Give a brief list of requirements for good basement lighting.
16. Where would 3-way switches be installed in a basement lighting system?
17. How is a garage lighting system planned? What is the most important requirement?
18. Name a special wiring installation that may be used in a garage.
19. Explain the operation of a master control switch.
20. Why would an electrician avoid using oversize wire with a given outlet or switch?
21. How does a glow-type starter differ from a manual starting switch?

CHAPTER 13

Electric Heating

Practical electricians are called upon to make many types of electric heating installations. *Space heaters* are in most common use. A space heater is defined as a heater without a reflector or other device for directing the heat output; the heat is produced by electric current flow through resistive elements and is distributed into the surrounding space chiefly by convection and conduction. Transfer of heat by *conduction* means the flow of heat through a solid substance such as iron. Transfer of heat by *convection* means the carrying of heat as by air rising from a heated surface. Both conduction and convection are different from *radiation* of heat, because *radiation* takes place in the absence of matter, as in the passage of heat through the vacuum inside the bulb of an incandescent lamp.

Electricians work with two types of space heaters: (1) the steel-sheath type which can be operated up to 750°F; (2) the porcelain-enameled type which can be operated up to 1200°F. Space heaters with alloy-steel sheaths are similar to the porcelain-enameled types in that they can also be operated up to 1200°F. Both of the basic types of heaters are used in appliances such as ovens. In room heating, both convection and radiation provide distribution of heat into the sur-

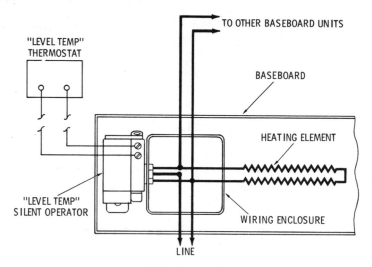

(A) Baseboard unit.

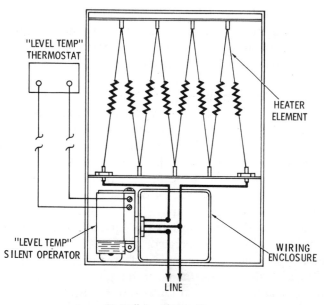

(B) Wall or ceiling unit.

Fig. 1. Various types of

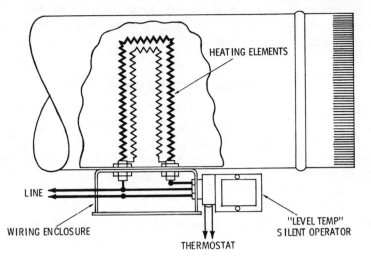

HEATING ELEMENTS

LINE

WIRING ENCLOSURE

THERMOSTAT

"LEVEL TEMP"
SILENT OPERATOR

(C) Electric duct unit.

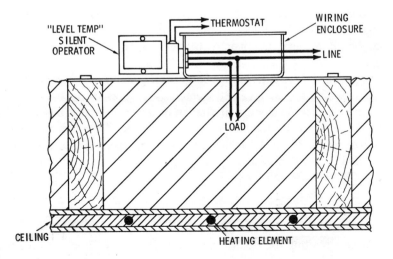

"LEVEL TEMP"
SILENT
OPERATOR

THERMOSTAT

WIRING
ENCLOSURE

LINE

LOAD

CEILING

HEATING ELEMENT

(D) Ceiling cable unit.

electric heating units.

rounding space. As a rough rule of thumb, from 1 to 2 watts per cubic foot is required to heat the air in a room when the temperature is near the freezing point outside.

GENERAL INSTALLATION CONSIDERATIONS

Electric heaters may be built into walls or ceilings, or may be mounted along the baseboard. Switching facilities are usually automatic and include a thermostat. Typical units are shown in Fig. 1. A wall heater may be provided with one or more fans to provide forced-air circulation. This feature gives more rapid distribution of warm air when the heater is first turned on. A typical wall heater of this type fits in a 7¼ × 14 inch wall opening, and consumes 1500 watts. Larger wall-heater units consume up to 4000 watts. In many designs, fans can be operated with the heater element turned off so that air circulation is provided in summer.

Baseboard heaters are manufactured in lengths from 28 to 107 inches, and consume from 300 to 2000 watts. In some installations, an electrician may use portable baseboard heaters. They are provided with carrying handles and are plugged into convenience outlets. A portable baseboard heater is often desired for use in sun porches and other seasonally occupied rooms. This type of heater is manufactured in lengths from 27 to 71 inches and consumes from 300 to 1500 watts. Baseboard heaters are usually between 9 and 10 inches high and between 3 and 4 inches deep.

Fig. 2 shows the exterior appearance of permanent baseboard, portable baseboard, and recessed heaters. A recessed electric heater is basically a drop-in floor unit that may be used where baseboard or wall heaters are unsuitable. Lengths of recessed heaters range from approximately 35 to 107 inches and consume from 300 to 2000 watts. Baseboard and recessed heaters generally operate by convection and do not have forced-air fans.

CIRCUITING

Basic circuiting for electric heaters is shown in Fig. 1. A bimetallic thermostat is used to turn the heater on automatically when the room temperature falls below a preset level. Similarly, the thermostat turns

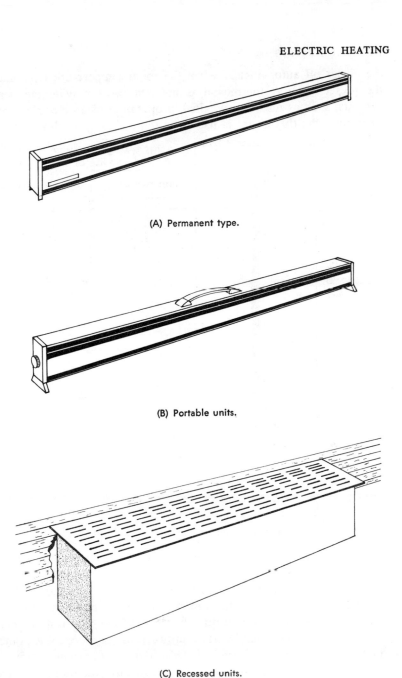

(A) Permanent type.

(B) Portable units.

(C) Recessed units.

Fig. 2. Baseboard electric heating units.

the heater off automatically when the room temperature rises above the preset level. A thermostat is not suitable for switching heavy currents, such as must be switched in operation of an electric heater. Therefore, the thermostat is used in a relay circuit, as shown in Fig.

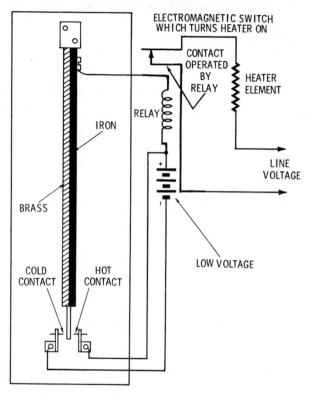

Fig. 3. Compound metal thermostat and circuit.

3. Thus, a small current in the thermostat branch can switch a heavy current in the heater branch.

The thermostat cannot switch the heater circuit on or off directly because the thermostat contacts move together or move apart slowly. Arcing at the contacts would quickly burn out the thermostat. Therefore, low voltage and small current are used in the thermostat branch. A current of 0.2 ampere is typical. This thermostat current is provided by a small voltage-stepdown transformer. In turn, the thermostat current operates a power relay, commonly called an *operator*.

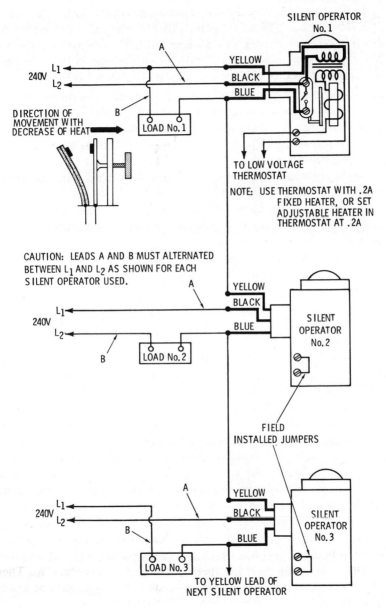

SILENT OPERATOR
No. 1

240V

L_1

L_2

A

YELLOW

BLACK

BLUE

B

LOAD No. 1

DIRECTION OF
MOVEMENT WITH
DECREASE OF HEAT

TO LOW VOLTAGE
THERMOSTAT

NOTE: USE THERMOSTAT WITH .2A
FIXED HEATER, OR SET
ADJUSTABLE HEATER IN
THERMOSTAT AT .2A

CAUTION: LEADS A AND B MUST ALTERNATED
BETWEEN L_1 AND L_2 AS SHOWN FOR EACH
SILENT OPERATOR USED.

A

YELLOW

BLACK

BLUE

240V

L_1

L_2

B

LOAD No. 2

SILENT
OPERATOR
No. 2

FIELD
INSTALLED JUMPERS

A

YELLOW

BLACK

BLUE

240V

L_1

L_2

B

LOAD No. 3

SILENT
OPERATOR
No. 3

TO YELLOW LEAD OF
NEXT SILENT OPERATOR

Courtesy White-Rodgers Co.

Fig. 4. A sequence system for one low-voltage relay and three thermal-type power relays.

An operator also contains a thermostatic device called a *thermal-type relay*. It has a bimetal heater which is warped when heated by a resistive element connected in series with the thermostat. In turn, warping of the bimetal blade suddenly trips a snap switch and closes the power circuit to the heater. The snap switch permits the power circuit to be opened or closed quickly, thereby minimizing arcing at the contacts. Thus, a small current in the thermostat branch effectively controls a large current in the heater-element (load) branch.

Fig. 4 shows the circuiting details for three heater loads using one thermostat and three operators. This is a *sequence* system in which the loads are switched on in succession at 45-second intervals and are switched off successively at 45-second intervals. Sequence operation is desirable when several loads are to be switched because the thermostat carries only 0.2 ampere in a sequence system. If sequencing were not used, the thermostat would have to carry 1 ampere for controlling five loads—this would tend to shorten the life of the thermostat.

The sequence of operation in Fig. 4 is as follows: When the contacts of the low-voltage thermostat close, 0.2 ampere flows through the heating element in the operator (thermal-type relay). In approximately 45 seconds, the bimetal blade warps sufficiently to suddenly close the snap switch for load No. 1. At this time, the step-down transformer in the second operator is energized, and its thermal-type relay is tripped about 45 seconds later. As soon as the snap switch for load No. 2 is closed, the step-down transformer in the third operator is energized, and the snap switch for load No. 3 closes about 45 seconds later. When the room comes up to the preset temperature, the low-voltage thermostat opens its circuit, and load No. 1 switches off after 45 seconds, followed by switch-off of load No. 2 and load No. 3 at 45-second intervals.

In some installations, an electrician may need to install a mechanical switch to supplement control by a thermostat. A *limit switch* is used for this purpose, as shown in Fig. 5. This type of switch is different from a manually-operated switch in that the switch is automatically operated by a moving object such as a cabinet door, which de-energizes the heater when the door is closed. Since the limit switch is connected in series with the heating elements, the heater cannot be energized by the thermostat as long as the door is closed.

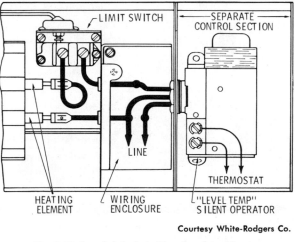

Fig. 5. Limit switch included in a baseboard heater.

AC OPERATED RELAYS

In the foregoing examples, the relay contacts were closed or opened by means of mechanical forces. Thermal-type relays use the mechanical force produced by warping of a bimetallic blade. Limit switches use mechanical force from a moving object. In addition to these devices, electricians may use relays that operate from electromagnetic force in an AC circuit. These are somewhat similar to relays used in DC circuits, except that special construction is employed in an AC relay to avoid armature vibration (chattering). In other words, a 60-Hz AC current rises to a peak and falls to zero 120 times a second. This vibration in electromagnetic force will cause a simple iron armature to chatter.

Chatter is objectionable, not only because it makes a loud buzzing sound, but also because of aggravated contact arcing. Therefore, *shading coils* are used in AC relays, as shown in Fig. 6. Another problem in AC relays is heating due to eddy-current losses in the magnetic circuit. Accordingly, the core of an AC relay is generally laminated, as explained in a previous chapter. In Fig. 6, the shading coil consists of a heavy copper loop. Current is induced in the loop by transformer action. Therefore, the electromagnetic forces on the armature are produced both by the relay-coil current and by the shading-coil current.

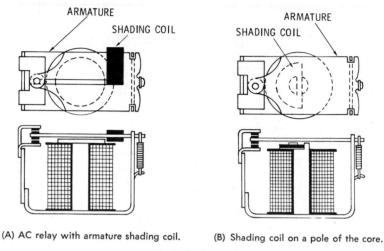

(A) AC relay with armature shading coil.   (B) Shading coil on a pole of the core.

Fig. 6. Shading coil.

We know that the current lags in an inductive circuit. In turn, although a single-phase current flows through the relay coil in Fig. 6, a current with lagging phase flows in the shading coil. Thus, the combined magnetic force produced by relay coil and the shading coil rises to a peak and falls to zero 240 times a second. Since the armature cannot vibrate to any appreciable extent at a rate of 240 times a second, the chatter is minimized and the AC relay operates in much the same manner as a DC relay. Vibration cannot be eliminated entirely, but it is reduced to a very small amount by the shading coil, which makes AC operation practical.

HOT-WATER ELECTRIC HEAT

Another type of electric space heating uses hot water without a plumbing installation. Fig. 7 shows the essential features of this heating method. An electrical element A inside the copper tubing heats a permanently sealed-in water and antifreeze solution. Operation of the electrical element is thermostatically controlled. Since hot water tends to rise, it circulates upward and through the finned heat-distribution area B where heat is transferred from the fin surfaces to the surrounding air. This cools the water, which causes it to circulate down and

376

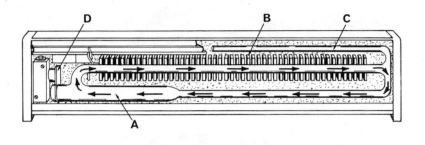

Courtesy International Oil Burner Co.

Fig. 7. Basic function of a hot-water electric heater.

back over the electrical element. Expansion space is provided by chamber C.

In case the water becomes overheated for any reason, the enclosure cannot explode due to limit switch D. As the temperature of the copper tube increases, the tube expands and increases in length. When a certain length is exceeded, the limit switch is tripped which opens the circuit of the electrical element. Antifreeze is used in the water to avoid damage due to freezing in case the unit is exposed to low winter temperatures.

RADIANT HEATER

A radiant heater is defined as an electric heating unit which has an exposed incandescent heating element. The electrical element glows red in operation, and a reflector is provided which directs heat radiation out through the grille much as a mirror reflects light. Heat reflectors are generally made from bright metal. It is basically the reflector which distinguishes a radiant heater from a space heater. Fig. 8 shows a small radiant heater, such as installed in bathrooms. This is a ceiling-type heater which contains a fan to supplement heat radiation by convection. Units of this type consume 1250 watts.

Radiant heaters may also be installed in a wall. If an exhaust fan is provided, the heater can also serve as a ventilator in summer. Radiant heaters mounted in a wall are suitable for either kitchen or bathrooms. They operate from either 117-volt or 234-volt lines, and a typical

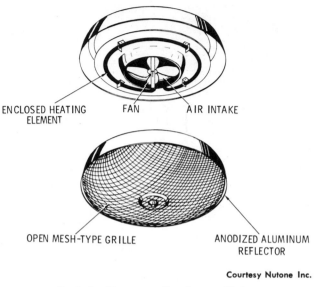

ENCLOSED HEATING
ELEMENT

FAN

AIR INTAKE

OPEN MESH-TYPE GRILLE

ANODIZED ALUMINUM
REFLECTOR

Courtesy Nutone Inc.

Fig. 8. A ceiling-type radiant heater with fan.

heater consumes 1600 watts. A radiant wall heater is usually provided with a manual switch but may also be installed with an automatic thermostat control if desired. Thermostats operate as previously explained for space heaters.

ELECTRIC WATER HEATERS

Electric water heaters use *immersion heaters* which are designed to operate in water. Fig. 9 shows some typical immersion heater elements. A snap-action thermostat is generally used with an immersion heater, as shown in Fig. 10. The snap action provides sudden opening or closing of the contacts, thereby minimizing arcing at the contacts. Installation of the pipes and the heater is a plumber's job; the electrician is concerned only with installation of the electrical system.

Since hot water tends to rise, a tall heater may be manufactured with one immersion heater at the bottom and another at the top. In such case, the upper heater is often operated by a double-throw snap-action thermostat, as shown in Fig. 11. If the lower heater unit consumes 1000 watts, the upper heater unit typically consumes 1500

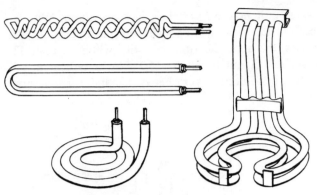

Fig. 9. Electric heating elements for water heaters.

watts. Each of the heater units is separately connected to the line through fuses or circuit-breakers in a connection box. Note in Fig. 11 that a double-throw thermostat energizes the upper heater first. After the water in the top of the tank comes up to the preset temperature, the left-hand contacts are closed so that the lower heater unit can be energized. In turn, the water at the bottom of the tank comes up to the preset temperature.

The use of two heater units provides a better supply of hot water under conditions of varying demand. With a small demand, only the

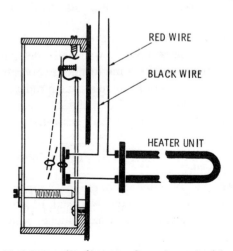

RED WIRE

BLACK WIRE

HEATER UNIT

Fig. 10. A snap-action thermostat for an immersion heater.

379

lower heater unit may be switched into operation. With a large demand, both heater units will be switched into operation; after the demand ceases, the upper heating unit will switch off first. Thermostats may be set to cut out at about 150°F in a typical installation. Of course, under conditions of very heavy demand, the temperature of the hot water may drop.

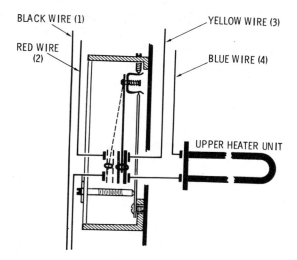

RED WIRE (2)

BLACK WIRE (1)

YELLOW WIRE (3)

BLUE WIRE (4)

UPPER HEATER UNIT

Fig. 11. A double-throw snap-action thermostat.

Before making an installation, the electrician should check applicable codes. For example, a separate watt-hour meter may be required for an electric water heater. In some locations, operation may be forbidden during certain hours, and the electrician must install an electric time switch to open the heater circuit during these hours Only the reserve supply of hot water in the tank is available while the heater circuit is open. Hence, a comparatively large tank may be used, or a small supplementary tank which does not require an electric time switch may be installed.

Time switches may be connected on the supply side of the disconnecting means, according to the National Electrical Code. Taps from service conductors to supply time switches may be installed as separate conductors, in cables approved for the purpose, or enclosed in rigid conduit; electrical metallic tubing may also be used. The service-entrance conductors must not be run within the hollow spaces of frame buildings unless fuses or circuit breakers are installed at the outer end of the conductors.

Means must also be provided for disconnecting the conductors from the service-entrance conductors. (See Fig. 12.) This provision must

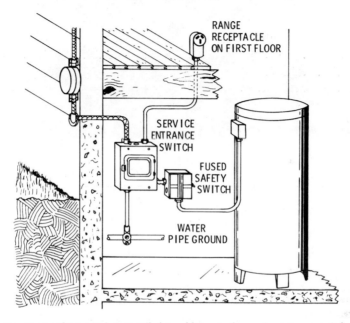

Fig. 12. A safety switch is installed in addition to the service-entrance switch.

be located at a readily accessible point nearest the entrance of the conductors, either inside or outside of the building wall. An approved manually-operated switch may be used, or a circuit breaker of the air-break or oil-immersed type with a handle which is marked and identified. A typical circuit breaker with an external handle is shown in Fig. 13. An external handle is required. Where electrical remote control is desired, a push-button control can be used in addition to the manual handle.

Note that a common enclosure, or a group of separate enclosures, must not contain more than six switches or six circuit breakers. Two or three single-pole switches or breakers, capable of individual operation, may be installed on multiwire circuits, with one pole for each un-

Fig. 13. A single-pole circuit breaker.

grounded conductor, as a unitary multipole disconnect, *provided they are equipped with "handle ties"* or equivalent approved arrangement. If the circuit breaker does not open the grounded conductor, another switch must be provided in the service cabinet for disconnecting the grounded conductor from the interior wiring.

A unitary multipole disconnect switch has holes drilled through the individual switch handles through which a metal rod is passed. The ends of the rod are secured by a rectangular metal stamping which surrounds the handles. Thus, all of the individual switch handles are operated simultaneously and cannot be operated separately. This simultaneous disconnect installation is required for all two-phase or three-phase circuits, but is not required for single-phase circuits.

As seen in Fig. 13, circuit breakers are basically electromagnetic relays which open automatically when the current demand is excessive. This is the most common type of breaker, and it must be reset manually after it has been tripped due to an overload. A breaker will immediately trip again in case the cause of the overload has not been corrected. When new equipment is used in an installation, the most likely cause of overload is an error in wiring connections. In case of difficulty, it is advisable to have the connections checked by an assistant in order to get new ideas in following the wiring diagram. Automatic fuseless circuit breakers are available in ratings up to 200 amperes and in 10 different box sizes. A 2-circuit service panel is used with water heaters. For general-purpose installations, 12-, 14-, or 24-circuit service panels may be used. Circuit breakers are separate from the panels and are plugged in; never install a circuit breaker with a rating higher than the allowable current-carrying capacity of the wires used in its circuit. There are six breaker ratings in common use; 15- or 20-ampere single-pole breakers are used for general lighting installations. For heavy loads, 20-, 30-, 40-, or 50-ampere double-pole breakers are used.

GROUND CIRCUITS

The National Electrical Code requires that metal frames of water heaters operating on circuits above 150 volts to ground shall have a ground circuit as shown in Fig. 12. If grounding happens to be impractical in a particular installation, special permission must be obtained, and the metal frames must then be permanently and effectively insulated from the ground. It is recommended that frames be grounded, even on circuits operating at less than 150 volts above ground This is always good practice because 117 volts can give a person a fatal shock.

The white (neutral) wire of all AC installations must be grounded. A No. 6 or No. 4 copper ground wire is the size normally used. If No. 8 wire is used, it must be armored as shown in Fig. 14. Larger sizes of ground wires need not be armored provided there is no danger of mechanical damage. Note that in a rural installation the ground wire does not run through the entrance switch, but is tapped off the neutral overhead wire, and brought down the side of the house or yardpole, and connected to a ground rod or to an underground water-pipe system (see Fig. 14).

If a ground rod must be used, either a copper rod or a galvanized iron or steel pipe is suitable. A copper ground rod must be at least ½ inch in diameter and a galvanized pipe at least ¾ inch in diameter according to most codes. The rod or pipe must be at least 8 feet long, located at least 2 feet from any building, and the top driven at least one foot below the surface. Thus, the top of the rod and the connection to the ground clamp are buried below the surface.

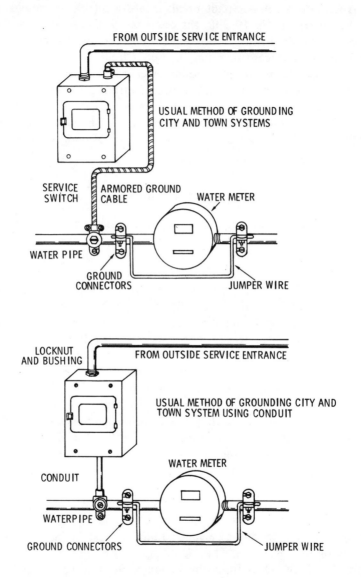

FROM OUTSIDE SERVICE ENTRANCE

USUAL METHOD OF GROUNDING
CITY AND TOWN SYSTEMS

SERVICE
SWITCH

ARMORED GROUND
CABLE

WATER METER

WATER PIPE

GROUND
CONNECTORS

JUMPER WIRE

LOCKNUT
AND BUSHING

FROM OUTSIDE SERVICE ENTRANCE

USUAL METHOD OF GROUNDING CITY AND
TOWN SYSTEM USING CONDUIT

WATER METER

CONDUIT

WATERPIPE

GROUND CONNECTORS

JUMPER WIRE

Fig. 14. A ground rod is used if

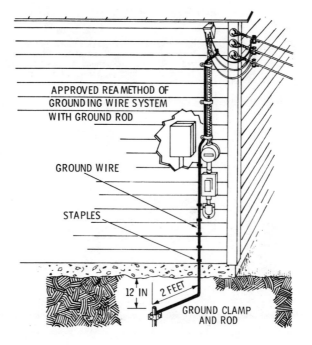

APPROVED REA METHOD OF
GROUNDING WIRE SYSTEM
WITH GROUND ROD

GROUND WIRE

STAPLES

12 IN   2 FEET

GROUND CLAMP
AND ROD

a water-pipe ground is unavailable.

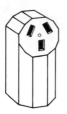

Cord sets are available in different lengths for connection to high-speed dryers, ranges, etc. Permits easy disconnection of equipment when redecorating, cleaning and servicing. Connect with 3 wire cable...For equipment using 240 volts - 50 amperes or less.

(A) Surface-type receptacle.

Cord sets are available to complete the connection to standard dryers, etc., to permit easy disconnecting of dryer for cleaning and servicing. Connect with 3 wire cable...For equipment using 240 volts - 30 amperes or less.

(B) Surface-type receptacle with L shaped ground.

Also available in duplex, but some areas will not permit a duplex receptacle of this type. For standard switch box and receptacle plate......For equipment using 240 volts - 15 amperes or less such as small air-conditioners, etc.

(C) Single receptacle with crow-foot blade.

Fits any standard switch box and uses a standard single receptacle plate. Connect with 3 wire cable ...For use with equipment using 240 volts - 20 amperes or less, such as larger air conditioners, power tools, garden equipment, etc.

(D) Single receptacle with tandem blades.

Fig. 15. Typical polarized receptacle.

There is an increasing trend toward the use of *polarized* devices with any type of electric heater or similar appliance. A polarized device has two current-carrying contacts plus one grounding contact. Polarized devices guard against dangers from current leakage due to faulty insulation or exposed wiring and help to prevent accidental shock. Typical polarized receptacles are shown in Fig 15.

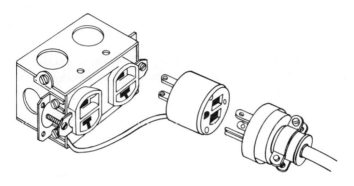

Fig. 16. Adapter used to provide a ground circuit to a standard receptacle.

If an existing installation has standard receptacles, polarized plugs are used with an adapter, as shown in Fig 16. In case the wires which connect the outlet to the service are run in conduit or armored cable, the lug on the lead wire of the adapter is connected to one of the screws holding the receptacle to the box. If the wires to the receptacle are run in nonmetallic cable with a bare ground wire, the lug on the lead wire of the adapter is connected to this bare wire. However, if the wires to the receptacle are in nonmetallic cable without a bare ground wire, the lug on the lead wire of the adapter must be specially connected to an approved ground such as a water pipe.

HEATING CABLES

A special type of electric heater is manufactured in cable form, as shown in Fig. 17. Heater cable is available in lengths from 20 to 60 feet. It has a flexible lead-covered construction. This type of electric heating is used to prevent frozen water pipes, to keep plants from being damaged by cold, in gutters, troughs, and animal drinkers. Heater

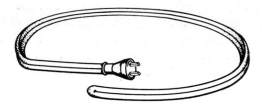

Fig. 17. A short length of heater cable.

cables may be used under porches, garage driveways, and sidewalks to keep ice and snow melted. Its lead-covered construction makes the cable weatherproof.

SUMMARY

Three types of heat transfer are conduction, convection, and radiation. Conduction means the flow of heat through a solid substance such as iron. Transfer of heat by convection means the carrying of heat by air rising from a heated surface. Radiation takes place in the absence of matter, as in the passage of heat through the vacuum inside the bulb of an incandescent lamp.

Electric heating may be built into walls or ceilings, or may be mounted along the baseboard. A recessed electric heater is basically a drop-in floor unit that may be used where baseboard or wall heaters are unsuitable. Baseboard and recessed heaters generally operate by convection and do not have forced-air fans.

A bimetallic thermostat is used to turn the heaters on automatically when the room temperature falls below a preset level. Thermostats are not suitable for switching heavy currents, such as must be switched in the operation of an electric heater. Thermostats are used in a relay circuit where small currents can operate branch circuits to switch heavy current in the heater units.

Another type of electric space heating uses hot water without plumbing installations. An electrical element inside the copper tubing heats a permanently sealed solution. When heated, the solution tends to expand, sending the water through the finned-head distribution area, where it is transferred from the fin surface to the surrounding air. This action also cools the water which causes it to circulate down and back over the electrical heating elements.

Radiant heating is defined as an electric heating unit which has an exposed incandescent heating element. The electrical element glows red in operation, and a reflector is provided which directs heat radiation out through a grille much as a mirror reflects light.

Radiant heaters may also be installed in a wall. If an exhaust fan is provided, the heater can also serve as a ventilator in the summer. Radiant wall heaters are usually provided with a manual switch but may be installed with an automatic thermostat control if desired.

Electric water heaters use immersion type heater elements which are designed to operate in water. Since hot water tends to rise, some manufacturers use only one element at the top and one element at the bottom. With a small demand, only the lower heater unit may be used, while with a large demand both heater units may operate together.

TEST QUESTIONS

1. Define a space heater.
2. How does heating by convection differ from heating by conduction? By radiation?
3. What is the difference between a heater thermostat and a heater operator?
4. Explain the operation of a thermal-type relay.
5. Why is a sequence system preferred for operation of several heater loads?
6. Discuss the principle of a limit switch.
7. How does a shading coil function in an AC relay?
8. Why are laminated cores used in AC relays?
9. Describe the operating cycle of a hot-water electric heater.
10. Explain the features of a radiant heater.
11. How does a snap-action thermostat operate?
12. When is a double-throw snap-action thermostat used in an electric water heater?
13. Define an immersion heater.
14. What is an electric time switch? Why might this device be required in an electric water-heater installation?

15. How can an electrician determine whether a separate watt-hour meter is required with an electric water heater?
16. Where should a fused safety switch for an electric water heater be installed?
17. Why must metal frames of electric water heaters always be grounded?
18. How does an electrician provide a ground connection in rural areas?
19. Define a polarized electrical device.
20. Why are polarized plugs and receptacles being used extensively for all appliances?
21. How is a polarized adapter connected to a standard receptacle?
22. Discuss several types of polarized receptacles.
23. What is the function of heating cable?
24. Where does an electrician usually install heating cable?
25. Why is heating cable enclosed in a lead sheath?

CHAPTER 14

Intercommunication
and Alarm Installations

An intercommunication (intercom) system provides speaker communication between two or more locations without the use of hand microphones. Fig. 1 shows the appearance of a typical intercom set. A single amplifier using vacuum tubes or transistors is generally employed in an intercom set to amplify the voice signal so that it can be heard on a speaker at the receiving end. Thus, an intercom system is basically a simplified and private telephone system.

Fig. 2 shows the circuit for a vacuum-tube intercom set. A two-stage audio amplifier is used in this example, comprising a 6AV6 and a 6EH5 tube. The 6X4 tube is a rectifier which changes the 117-volt AC supply into pulsating DC. The pulsating DC is changed into a smooth DC supply voltage by the filter comprising C_4, R_8, and C_5. Potentiometer R_1 is a volume control. SW_1 is an off-on power switch, and SW_2 is a talk-listen switch. S_1 is a local speaker, which is part of the intercommunciation set. SW_3 is a station-selector switch. S_2 and S_3 are remote speakers connected by lines to the intercom set. Any desired number of remote speakers can be wired into the system.

Note in Fig. 2 that the talk-listen switch is set to the "listen" position for the intercom set location. The station-selector switch is set for remote unit 1. Thus, S_2 operates as a microphone, and when a person speaks into it, the voice signal is amplified and reproduced by S_1. If the operator at the S_1 location wishes to speak to the S_2 location, he turns the talk-listen switch to the "talk" position. In turn, S_1 is connected to operate as a microphone. The voice signal from S_1 is amplified and reproduced by S_2. Again, if communication is desired between S_1 and S_3, the station-selector switch is turned to position 2.

In this type of intercom system, the operator at the S_1 position has complete control of the system operation. Whether S_2 or S_3 locations

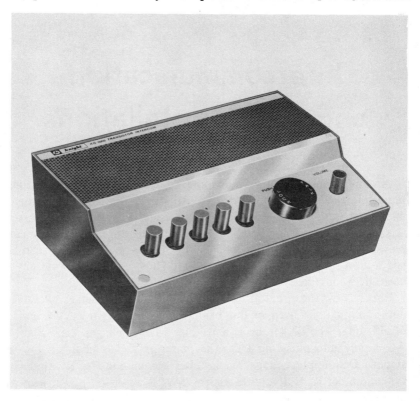

Courtesy Allied Radio Corp.

Fig. 1. A typical intercom unit.

are operative depends on the setting of SW_3. Similarly, whether the intercom set is operating as a transmitter or as a receiver depends on the setting of SW_2. Therefore, if it is desired to elaborate the system so that calls can be made at any time from the remote locations, additional intercom sets must be wired into the system, or another type of equipment must be used which provides full two-way operation.

392

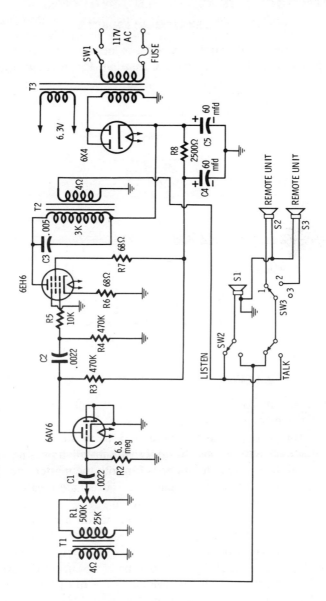

Fig. 2. Intercom set with master unit and two or more remote units.

SYSTEM PLANNING

The station at which the amplifier is located is called the *master* and each remote station is called a *substation*. Fig. 3 shows the plan of a

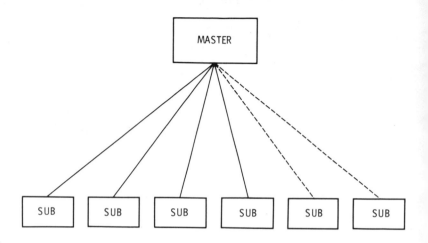

Fig. 3. System employing one master and any number of subs.

master station connected to a number of substations, while Fig. 4 shows the plan of a complete master installation for all stations. In other arrangements, some stations may be master installations, while the others are substations. In *nonprivate* installations, the master station's push-to-talk switch functions for both the master and the substation. Thus, only the master location can originate a call, and in addition, can listen-in on a selected substation.

A nonprivate installation is commonly used for a front-door station. When the door bell rings, the master station can be switched to the front porch to carry on a conversation with a visitor before opening the door. A home intercom system may be elaborated to operate also as a music distribution system to various rooms. In such case, a radio tuner and/or record player will be added, as illustrated in Fig. 5. These elaborated systems employ better-quality speakers for improved music fidelity and a switching function at each substation to permit selection

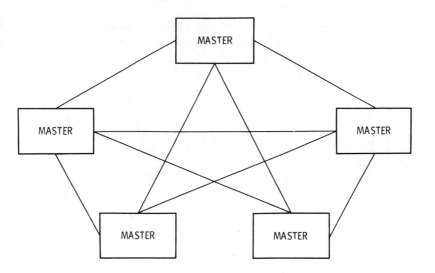

Fig. 4. Intercom system using master units at each sub.

of a particular music channel. Deluxe systems even include a dual-speaker installation in each room for stereo reproduction of music.

You must first familiarize yourself with the equipment that you plan to install, and then discuss its details with the office manager or home-owner. After determining the number and location of stations, a floor plan is drawn up, such as shown in Fig. 6. This is a plan for a sales office and stock room; however, the same type of plan is drawn up for a residence, apartment building, or rural installation. Keep in mind that the master station will need a source of AC power (unless a bat-tery-operated transistor system is used).

The National Electrical Code requires that interconnecting cable be run not closer than 2 inches to any AC line, even if the line is enclosed in conduit. This is not only a safety precaution but helps to minimize AC hum pickup. Interconnecting cables must not be run close to tele-phone wires to avoid pick-up of telephone signals. Interference due to stray-field pick-up can be minimized by using intercom sets that work into low-impedance lines. The circuit shown in Fig. 2 is an example of a low-impedance system.

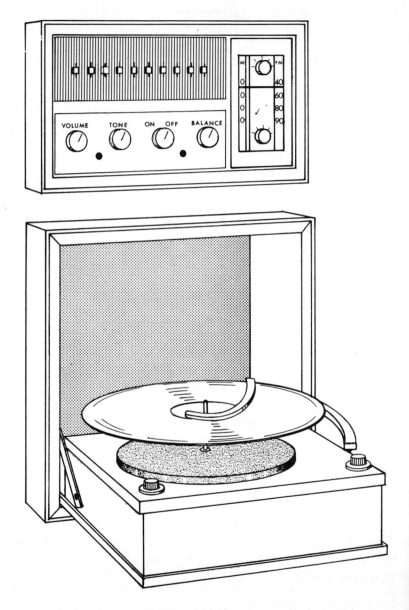

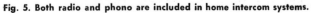

Fig. 5. Both radio and phono are included in home intercom systems.

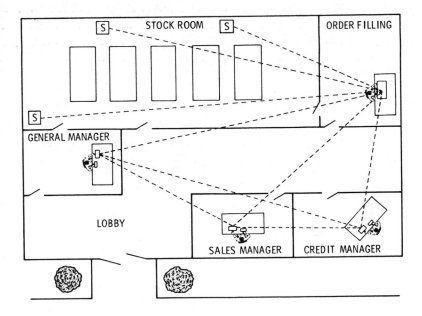

Fig. 6. Floor plan showing position of stations and communication between each.

There are other various factors to be considered in the layout of an intercom system. For example, the electrician must observe the construction of the building or house in which the installation is to be made. A brick house may have two courses of brick and only ¾-inch furring strips between the inner layer of brick and the plaster wall, which gives little space through which to draw cable. Only the inside walls, constructed of 2″ × 4″ studs have adequate clearance for cables. A brick veneer house has one course of brick, with 2″ × 4″ studs in the outer as well as the inner walls. Frame houses have 2″ × 4″ studs in all the walls.

Connecting an intercom system from room-to-room requires running the cables up the space between studs to the attic, across the attic, and then down between the studs in another wall. Another method is to run the cable down into the basement or crawl space, across and up into the wall between the studs. The bottom of the studs are usually nailed to 2″ × 4″ sole plates, which in turn are nailed to the flooring. Therefore, the electrician must drill through both the sole plate and the flooring to get into the space between studs. Similar construction is used at the top of the studs so that about the same amount of drilling is necessary to go up to the attic and over.

The most difficult part of the installation is running the cable from floor to floor in a two-story house. This kind of installation requires removing the baseboard on the second floor and drilling down to the space between the studs in the room below. Outlet holes to the stations should be drilled through the baseboard. A suitable snake is usually required to draw the cables. After the snake has been run through the space for the intercom wires, the wires are twisted through the hook and secured with tape. Then, the snake is pulled out, drawing the wires with it.

Installation is much easier in homes or buildings under construction. If junction boxes are used, leave about 3 inches of intercom cable

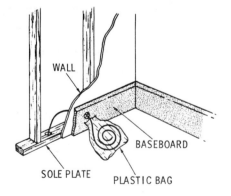

Fig. 7. Plastic bag protects cable.

hanging out for final connection. On the other hand, if the cable is to be run directly to the stations, it is necessary to allow sufficient length according to the floor plan. Roll up the excess cable close to where it comes out of the baseboard and enclose it in a plastic bag, as shown in Fig. 7. This precaution keeps the cable clean during subsequent plastering and painting.

PLANNING DETAILS

It is necessary to determine whether the complete intercom installation is to be in-wall or on-wall, or whether only the master station is to be built-in with the substations on-wall. For built-in installations in new homes, the electrician must discuss the plan with the plasterer so that he will know where the wall outlets are to be located. Otherwise, the electrician must return to make the cutouts himself after the plastering is completed and before painting is started. Most in-wall intercom stations hang from the plaster or wallboard, but some of the heavier units will require wood framing for support. Some intercom systems use metal boxes that can be put in by the plasterer.

If a front-door station is to be part of an intercom system installed in a brick house, it must be put in when the bricks are laid. Therefore, the architect should include the box in his masonry plans. When a single in-wall master amplifier is centrally located and is not a part of the master station, the electrician must provide an AC outlet where the amplifier is to be located. This should be a behind-the-wall outlet for a permanent type of connection. The same consideration also applies to in-wall master stations with a built-in amplifier.

Since the signal voltages from intercom amplifiers are low, it is usually adequate to install open wiring. However, on long runs that pass through strong stray fields, such as in heavy manufacturing areas, a conduit installation may be required to minimize noise interference. The basic intercom cable used between a master station and a substation has three conductors. In many systems, one wire is a common line, with the second wire used for incoming signals and the third wire used for outgoing signals. In any case, the electrician should check the installation instructions provided with the intercom system to determine the number of conductors that will be required.

For installation convenience, most intercom systems are designed to use multiconductor cable that can serve a number of stations. Some intercom systems are designed for use with *balanced* lines and use *twisted pairs*. Note that neither side of a balanced line is grounded. Fig. 8 shows examples of balanced and unbalanced line connections.

399

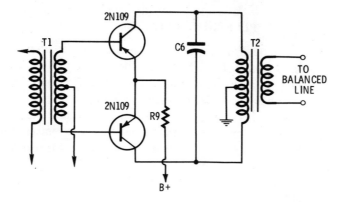

(A) Balanced.

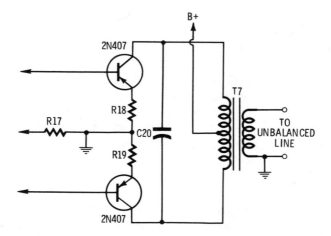

(B) Unbalanced.

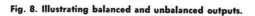

Fig. 8. Illustrating balanced and unbalanced outputs.

Other intercom systems used shielded wire for the incoming lead; however, most systems do not use shielded wire. Unless an electrician has had previous experience with a particular system, it is advisable to use the cable specified by the manufacturer, even if it costs more than ordinary cable. This precaution will ensure against unsatisfactory installations.

A good intercom system has reserve power so that an additional speaker can usually be operated in parallel with the original speaker if desired. As explained in a previous chapter, maximum signal power transfer is obtained when the speaker load matches the source impedance. The impedance of the intercom line is seldom a matter for concern because the lines are comparatively short with respect to voice-signal wavelengths. In an ideal installation using long lines, in which maximum fidelity and maximum power transfer is desirable, both the source and load impedances should be the same as (should match) the line impedance. Coaxial cable is supplied in a wide range of impedances, from about 50-ohms to 200-ohms. Twisted cable is also available in a wide range of impedances.

WIRELESS INTERCOM SYSTEMS

A wireless intercom requires no cable runs between stations. The 117-volt wiring system is used to conduct the voice signals from one station to another. This is done by means of a comparatively high-

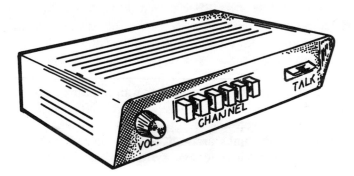

Fig. 9. A wireless intercom master unit.

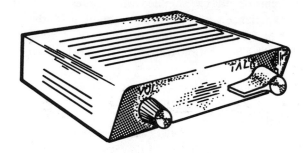

Fig. 10. A wireless intercom auxiliary unit.

frequency *carrier* current which can be easily separated from the low-frequency 117-volt current. A typical all-master wireless intercom system can use up to 12 master units (see Fig. 9). Each master can originate calls to any other master. It can also receive calls on any channel (carrier frequency). By switching several masters to the same channel, a conference system is provided.

A 12-station system can accommodate 6 separate conversations simultaneously. Auxiliary stations (Fig. 10) are used in locations where calls need not be originated, but replies must be made. An auxiliary station operates on one channel only, and hence provides communication with a single master location. Although a wireless intercom system is very easy to install, since each unit is merely plugged into a convenience outlet, the system also has certain disadvantages. The chief disadvantage is a tendency to reproduce line noises and hum. Interference can be minimized by careful adjustment of the signal filters in the units. However, complete elimination of all interference under various conditions of operation can seldom be obtained.

Wireless intercom systems occasionally pick up broadcast signals or code signals from nearby radio stations, with resulting low-level background interference. The interference level often changes when switches in the 117-volt line are opened or closed, and when various numbers of intercom units are switched into operation. Another disadvantage of a wireless intercom system is that communication is limited to the 117-volt circuit supplied by the local line transformer

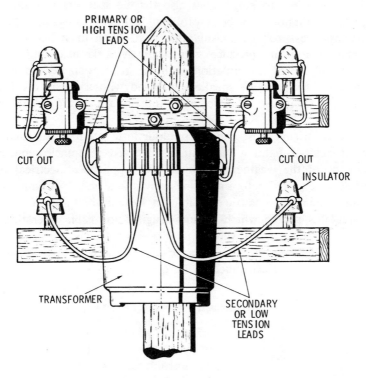

Fig. 11. A line transformer.

(Fig. 11). In other words, high-frequency carrier currents are stopped by a line transformer.

Privacy cannot be obtained with ordinary wireless intercom systems. If an apartment house is powered by a single line transformer, all occupants can hear any conversation with a suitable intercom unit. A wireless intercom system can be compared with a party-line telephone system in which a subscriber can listen-in on conversations between other subscribers. Although elaborate "speech-scramblers" can be used, these are quite expensive and are seldom installed.

MUSIC DISTRIBUTION SYSTEMS

A music distribution system is nearly always an in-wall installation which requires additional plaster cutouts, not only for the control amplifier/tuner/record player, but also for the speakers in each room. Proper ventilation may be a problem with in-wall systems because vacuum tubes produce considerable heat which must be allowed to escape. Transistors produce comparatively little heat and are advisable in case of a ventilation problem. It is often desired to have patio or other outdoor speakers installed with a music distribution system. In such case, suitable cables must be run to the outdoor locations.

The electrician must plan an outdoor installation in the same way as an outdoor lighting installation. Exposed receptacles and units must be of the weatherproof type. All applicable electrical codes should be checked before a wiring plan is finalized. If a home-owner wishes to install a music distribution system that uses a radio or high-fidelity source which is not designed for cable operation, the electrical requirements should be carefully drawn up by a highly experienced electrician or a specialist. Otherwise, the completed installation may leave much to be desired.

INDUSTRIAL INSTALLATIONS

Intercom installations in offices or plants are often easier than in homes. The manager will often permit the electrician to tack the cables on the outside of walls. This type of intercom wiring will usually meet electrical codes. However, the electrician must run the wires through the walls to get from one room or office to another. Either a star drill and hammer can be used, or an electric drill with a carbide-tipped bit. This part of an intercom installation follows the same methods used in a lighting installation, where cables are run up the space between the wall studs to the attic, across the attic, and then down between the studs in another wall.

When an installation is made in a plant, it is good practice for the manager and electrician to go over the entire area carefully during working hours. The noise levels can be noted, and in turn, an experienced electrician can give good advice concerning the number of stations that should be provided and the best location for each. This precaution is good insurance against misunderstandings that can result from installations which do not provide the intended utility and convenience for communication.

SCHOOL INSTALLATIONS

School intercom installations usually require a system with an "all-call" function, so that the principal can make a general announcement to all classes at the same time. This added function requires a few changes from ordinary cable runs, but is otherwise the same as home or small-office installations. Note that schools may have their own electrical codes which go beyond the requirements of codes applying to homes and offices. For example, intercom cables may have to be installed in conduit to meet a school code. Therefore, it is necessary to check with the school board before finalizing a school wiring plan.

The National Electrical Code provides that intercom conductors may be run in the same shaft with light and power conductors, provided that the intercom conductors are separated at least 2 inches, or where the conductors of either system are encased in noncombustible tubing. Conductors bunched together in a vertical shaft must have a fire-resistant covering capable of preventing fire from being carried from floor to floor, except where the conductors are encased in noncombustible tubing or are located in a fireproof shaft having fire stops at each floor.

BELLS AND CHIMES

Electric bells, chimes, or buzzers operate from comparatively low voltage, such as 6 volts. Either AC or DC may be used, although AC operation is most common. Fig. 12 shows a basic bell circuit using a

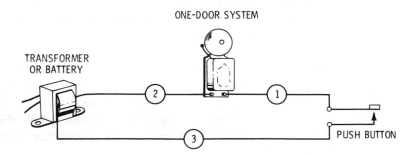

Fig. 12. An electric-bell circuit.

step-down transformer. Note that the primary of the transformer is permanently connected to the 117-volt AC supply line. The secondary circuit is closed by a push button to ring the bell. The primary has substantial inductance so that little idling current is drawn by the transformer. A bell-ringing transformer is designed to have considerable magnetic leakage so that it will not burn out in case of a short-circuit across the secondary. For example, if circuit points (2) and (3) should become short-circuited, there would be no danger of fire damage.

A door chime is wired in the same way as a bell. In the case of a two-chime unit, such as shown in Fig. 13, a three-wire installation is

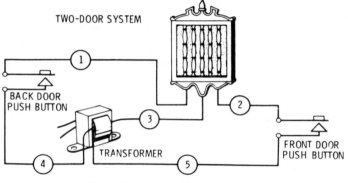

Fig. 13. A two-door chime circuit.

used. The transformer is the same as for a single bell or chime. However, the secondary terminals are connected to two branch circuits controlled by push buttons at individual doors. One branch circuit consists of conductors (2), (3), and (5); the other branch circuit consists of conductors (1), (3), and (4). Each chime has a distinctive tone to indicate which push button has been operated.

A tone signal system is shown in Fig. 14. This is the equivalent of a three-tone chime system, but employs a simpler wiring installation. The unijunction transistor operates in an oscillator circuit which supplies a tone signal to the speaker (or speakers). Note that a different value of resistance is connected in series with each push button. This results in a distinctive tone, depending on which button is operated. This system requires a DC supply, usually obtained from a transformer followed by a selenium rectifier and filter.

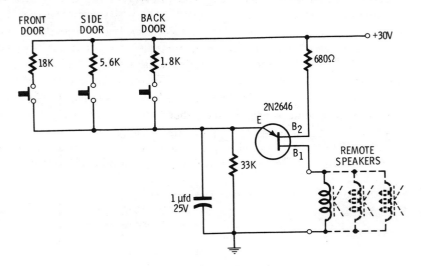

Fig. 14. A three-tone signal system.

Bells, buzzers, chimes, and signal systems must be installed in accordance with applicable electrical codes. Therefore, the electrician should check with local codes and with any special codes such as specified by schools and other public institutions. Hospital installations are comparatively complex and require careful planning. Special types of bells are used in outdoor installations, such as schoolyards, and are usually connected to electric timers.

PHOTOELECTRIC CONTROL OF BELL CIRCUITS

Electric bells installed in small business offices are often controlled by a photoelectric circuit, such as shown in Fig. 15. When a person passes through the doorway, a light beam is interrupted which causes the relay contacts to close. In turn, the bell circuit is closed and the bell (or chime) rings. The phototransistor is a device which changes light energy into electrical energy and produces amplified current flow in the relay circuit. When the light beam is interrupted, practically no current flows in the relay circuit. The potentiometer is a sensitivity control which is adjusted to obtain proper operation at the prevailing level of ambient light.

407

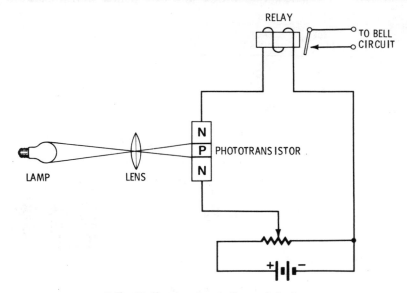

Fig. 15. Phototransistor bell-control circuit.

Phototransistors operate at comparatively low voltages. Although a battery is shown in Fig. 15 for simplicity, a commercial unit usually has a transformer, selenium rectifier, and filter to provide the DC supply. Installation of a photoelectric bell-control system must be made in such a way that excessive ambient light cannot strike the lens in front of the phototransistor. This often requires that a hood or tube be mounted in front of the lens so that most of the light which enters is provided only by the lamp.

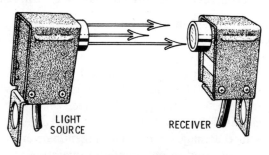

Fig. 16. An infrared alarm device.

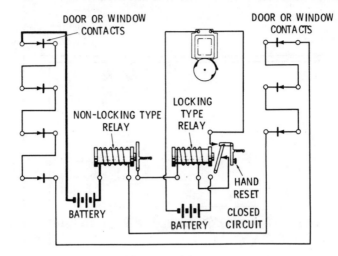

(A) Closed circuit.

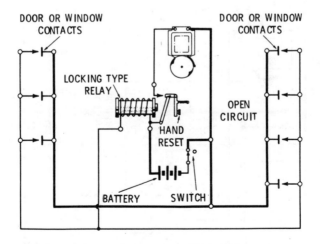

(B) Open circuit.

Fig. 17. Burglar alarm system.

409

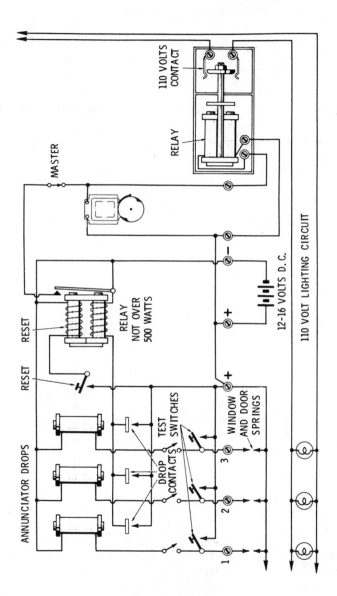

Fig. 18. Burglar alarm circuit having annunciator with relay.

ALARM DEVICES

Electricians often install various types of alarm devices. For example, most business establishments have a burglar-alarm system. Infrared light units (Fig. 16) are in wide use. These are similar to photoelectric bell-control units except that an infrared light source is used which is practically invisible. The receiver unit controls an electric bell which is usually mounted on the outside of the building so that it will attract the attention of police and passersby. Installation must be made in accordance with applicable electrical codes. Instructions for complete electrical connections are generally included with the equipment.

Other common alarm systems employ a wiring system with contacts that automatically open or close when a door or window is opened. These are classified into the closed-circuit and open-circuit systems, as shown in Fig. 17. The alarm bell may be in the building with the installation, or may be located at a remote point such as guardian station or police station. The more elaborate systems include annunciators to indicate the office, shop, or room that has been entered. A simple annunciator wiring system with relays to close lighting circuits, test switches, master switch, and bell is shown in Fig. 18.

Many offices and shops are equipped with fire- and/or smoke-alarm systems. These installations include a heat-sensor such as a thermostat or a number of thermostats located at suitable points, and connected to bells, annunciators, or both. Relays are used when the thermostat must switch a substantial amount of power. Smoke alarms are similar to the photoelectric systems used with doorbell installations. When the light beam is reduced in intensity by smoke, bells and/or annunciators are energized. The alarm indicators may be either locally or remotely located.

Photoelectric systems are also used to automatically switch night lights on and off in business institutions and factories. They are occasionally installed in residence wiring systems. Night lights are switched on after sunset and switched off after sunrise. Wiring instructions are included with the equipment, and the electrician must make the installation in accordance with the electrical code for that particular area.

SUMMARY

An intercom provides speaker communication between two or more persons at different locations. An intercom system is basically a simplified and private telephone system. The location which houses the amplifier is called the master unit and each remote location is called a substation. Each intercom system is designed to carry any number of substations, depending on individual requirements.

A nonprivate installation is commonly used for a front-door station. When the doorbell rings, the master station can be switched to the front door to carry on a conversation with the visitor before opening the door. A home installation can also be operated as a music distribution system to various rooms. In this case, a radio tuner and/or record player would be added. These elaborate systems use better quality speakers for improved music fidelity. Deluxe systems even include a dual speaker installation in each room for stereo reproduction of music.

There are various factors to be considered when laying out an intercom system. The installer must observe the construction of the building in which the installation is to be made. Connecting the intercom from room-to-room requires running the cable between wall studs, across the attic, and under floors. Installation is much easier in homes or buildings under construction.

It is necessary to determine whether the system is to be an in-wall or on-wall installation. If the front door is to be included in the system, a substation must be put in when the bricks are laid or wood siding installed. A good intercom system has a reserve power so that additional speakers can usually be operated in parallel with the original speakers.

A wireless intercom requires no cable runs between stations. The 117-volt wiring system is used to conduct the signal from one station to another. The chief disadvantage is a tendency to reproduce line noises and hum. Interference can be minimized by careful adjustment of the signal filters in the unit. Wireless intercom system also can pickup stray broadcast signals from nearby radio stations.

Various types of alarm devices are used in home and business. The most common unit is the infrared light beam. These units are similar to the photoelectric cell except that an infrared light source is practically invisible.

Other common alarm systems employ a wiring system with contacts that automatically open or close when a door or window is opened. Many offices and shops are equipped with fire- or smoke-alarm systems. When the light beam is reduced in intensity by smoke, bells are energized setting off the alarm system.

TEST QUESTIONS

1. Explain the function of an intercommunication system.
2. Why are amplifiers used in intercom systems?
3. What is the difference between a master station and a substation?
4. How does a music distribution system operate?
5. Discuss the difference between a private and a nonprivate intercom installation.
6. Explain how an AC power supply operates.
7. What type of cable is commonly used in an intercom installation?
8. Why should the electrician check with the plasterer before an installation is made in a new home?
9. When would an architect be concerned with details of an intercom installation?
10. How does a balanced line differ from an unbalanced line?
11. Describe the operation of a wireless intercom system.
12. What is the chief disadvantage of a wireless intercom system?
13. Why does a line transformer block intercom signals?
14. State some precautions to be observed in the installation of an industrial intercom system.
15. What special requirements may an electrician encounter in a school installation?
16. Explain the operation of the three-tone signal system.
17. How does a phototransistor bell-control system operate?
18. Why is a sensitivity control required in a photoelectric bell-control circuit?
19. Where would an infrared alarm device ordinarily be installed?
20. Explain the difference between a closed-circuit and an open-circuit alarm system.

21. What is an annunciator?
22. How does a fire-alarm system operate?
23. How does a smoke-alarm system operate?
24. Why are relays used in many alarm systems?
25. Explain how night lights are automatically switched on and off.

CHAPTER 15

Generating Stations and Substations

A *generating station* is a plant wherein electric energy is produced from some other form of energy by means of suitable apparatus. An *automatic station* is a station (usually unattended) which, under predetermined conditions, goes into operation by an automatic sequence and which maintains the required type of service; it goes out of operation by automatic sequence under other predetermined conditions and provides protection against operating emergencies. An automatic station may be either an automatic generating station or an automatic substation.

An electric power *substation* is an assembly of equipment for purposes other than generation or utilization, through which electricity passes for the purpose of switching or modifying its characteristics. Service equipment, distribution transformer installations, and other minor distribution equipment are not classified as substations. A substation is of such size that it includes one or more buses, circuit breakers, is usually a receiving point for more than one supply circuit, or sectionalizes the transmission circuits which pass through by means of circuit breakers.

GENERATING STATIONS

The plan of a generating station deals with the installation of generating units and all of the mechanical and auxiliary electrical equipment required to produce mechanical energy, change the mechanical energy into electrical energy, and deliver the electrical energy to the associated system. There are four classifications of generating stations:

1. Gas-power.
2. Oil-power.
3. Water-power (hydroelectric).
4. Steam generating.

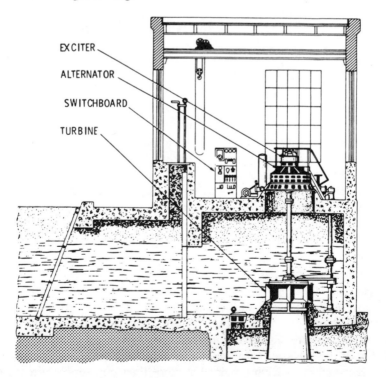

Fig. 1. A small hydroelectric generating station.

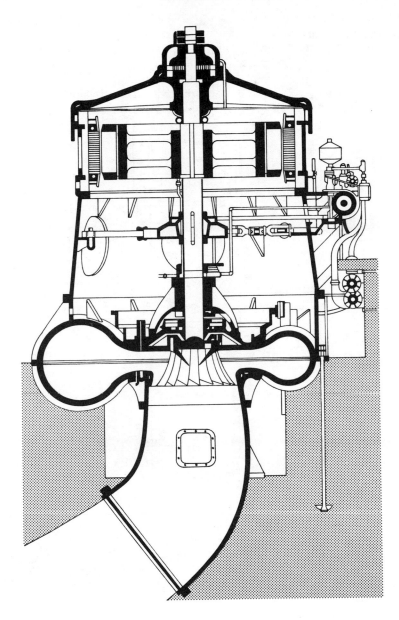

Fig. 2. Cutaway view of a typical generator.

417

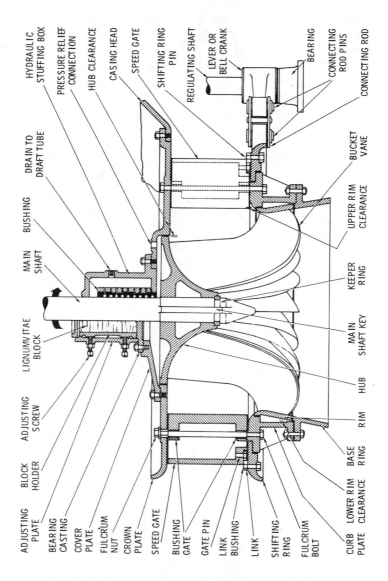

Fig. 3. Cutaway view of a water-reaction turbine.

However, the electrical features are quite similar in all four classes of stations. Fig. 1 illustrates a simplified section of a small hydroelectric generating station. Fig. 2 shows the plan of a generator and directly-connected turbine installation. The turbines, alternators, exciters, and controlling switchboard are housed in one large room of a hydroelectric generating station. Prime movers are arranged in a single row to simplify the penstock and tail-race design. The *turbine* is a rotary engine actuated by water under high pressure. An *exciter* is an auxiliary generator which supplies energy for the field excitation of another electric machine. A *prime mover* is an initial source of motive power which is applied to other machines. Fig. 3 shows a cutaway view of a water-reaction turbine.

The exciter voltage is 250 volts in large plants, and 125 volts for smaller installations. An exciter is generally a part of the generator assembly, as seen in Fig. 1. However, exciters driven by separate prime movers are also used. Generators are used that supply voltages up to 22,000 volts. Power outputs range up to 125,000 kilovolt-amperes (kva). Armature speeds range from 10 to 3600 RPM. Provisions are made to avoid damage in case of accidental short-circuits and to prevent overheating.

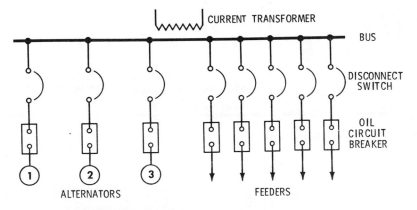

NOTE: BUS HAS A BAR CONSTRUCTION.
DISCONNECT SWITCHES ARE AIR SWITCHES.
FEEDERS CONNECT GENERATING STATIONS, OR CONNECT A GENERATING STATION AND A SUBSTATION.

Fig. 4. Single-bus, single circuit-breaker system.

Table 1. Typical Electrical Symbols

	MAIN CIRCUIT	CONTROL CIRCUIT
CONDUCTORS	NOT CONNECTED	CONNECTED
RESISTANCES		
	TERMINAL	GROUND
	FUSE	SHUNT
CONTACTS	OPEN	CLOSED
RESISTOR	TUBE TYPE	GRID TYPE
WINDINGS	SHUNT	SERIES
TRANSFORMER	AUTO	CURRENT
	POTENTIAL	POWER
AIR CIRCUIT BREAKERS	S.P.	D.P.
OIL CIRCUIT BREAKERS	S.P.	D.P.
INDUCTANCE		(REACTOR OR CHOKE COIL)

Single-Bus System

The main electrical connections in a generating station depend on the size and type of generator, the feeder arrangement, function of the generating station in the system, etc. Fig. 4 shows the simplest connection arrangement, called the *single-bus system*. This system is used only in small generating stations where the possibility of service interruptions is permissible. It provides adequate switching facilities and protection of apparatus in case of failure, but lacks flexibility. If any alternator circuit fails, the corresponding machine and circuit breaker must be withdrawn with its feeder circuits. The feeders are taken off

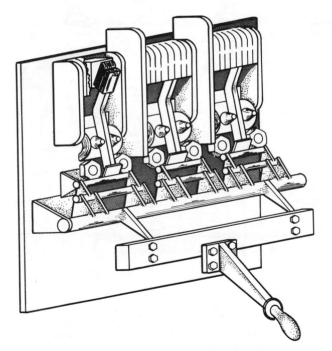

Fig. 5. Typical air circuit breaker.

between the alternators, and in case of insulation failure of a bus-bar support, a complete shutdown is necessary until the defect has been corrected. Table 1 shows typical electrical symbols.

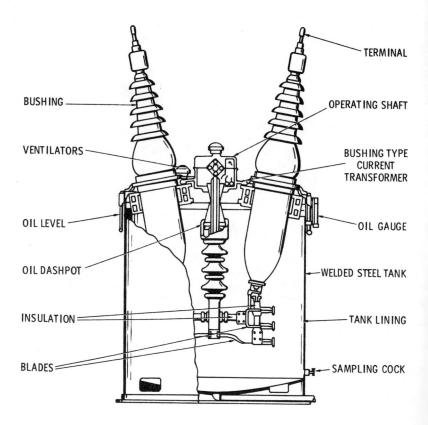

BUSHING

VENTILATORS

OIL LEVEL

OIL DASHPOT

INSULATION

BLADES

TERMINAL

OPERATING SHAFT

BUSHING TYPE
CURRENT
TRANSFORMER

OIL GAUGE

WELDED STEEL TANK

TANK LINING

SAMPLING COCK

(A) Oil.

Fig. 6. Small oil and

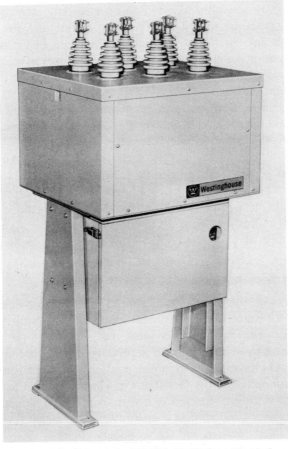

Courtesy Westinghouse Electric Corp.

(B) Vacuum.

vacuum circuit breakers.

Circuit Breakers

By use of sectionalizing disconnecting switches in the bus bars of a single-bus system, a complete shutdown of the station can be partly guarded against. A circuit breaker is a device for interrupting a circuit between separable contacts under normal or abnormal conditions. Circuit breakers are generally required to operate only infrequently, although some types are suitable for frequent operation. *Normal* operation means the interruption of currents not in excess of the rated continuous current of the circuit breaker. *Abnormal* operation means the interruption of currents in excess of rated values, such as caused by accidental short-circuits.

An oil circuit breaker has its contacts immersed in an oil bath. On the other hand, an air circuit breaker has its contacts exposed to surrounding air. A small air circuit breaker is shown in Fig. 5. The plan of a small oil circuit breaker is shown in Fig. 6A. Any circuit breaker must be designed to open the circuit rapidly, and to quickly "kill" the arc between its contacts. Elaborate means are provided to extinguish arcs in high-power circuit breakers. For example, a high-velocity blast of gas may be used to blow out an arc. Fig. 6B shows the appearance of a vacuum circuit breaker. The vacuum circuit breaker is a comparatively recent development.

Instrument Transformers

An *instrument transformer* is a special type of step-down transformer used for connection to a voltmeter or ammeter. A current transformer has its primary winding connected in series with the circuit in which current is to be measured. Its secondary winding is connected to an ammeter. On the other hand, a potential or voltage transformer has its primary connected across the circuit in which voltage is to be measured. Its secondary is connected to a voltmeter. Fig. 7 shows the connections for current and voltage transformers. Instrument transformers are used for safety reasons, and the secondary circuit is always grounded. It is very important to never open the secondary circuit of a "live" current transformer because of the high induced voltage.

Double-Bus System

To obtain greater flexibility and to reduce the possibility of service interruptions, two buses with additional disconnect switches or manu-

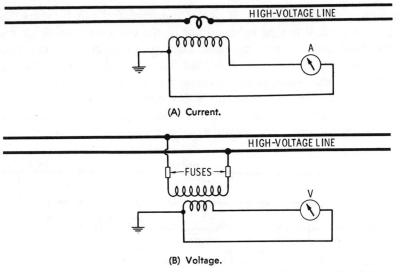

(A) Current.

(B) Voltage.

Fig. 7. Current and voltage transformer connections.

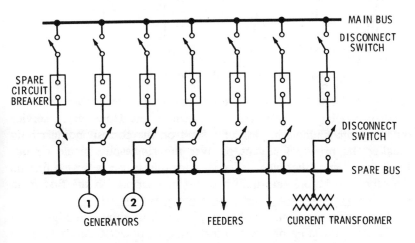

Fig. 8. Spare-bus, spare circuit-breaker system.

ally-operated oil circuit breakers may be used in small generating stations. One bus is called the main and the other is called the spare bus, as shown in Fig. 8. Ordinarily, the spare circuit breaker and spare bus are not used. However, if one of the regular circuit breakers develops a fault, it can be switched out of the circuit for repair or replacement, and the spare circuit breaker can be temporarily switched into the circuit. However, if a fault occurs in the bus insulation, complete station shutdown is necessary until the fault is corrected.

Another connection arrangement that employs two buses is shown in Fig. 9. This is called the double-bus, single-breaker system. It practically eliminates the possibility of a prolonged shutdown in case

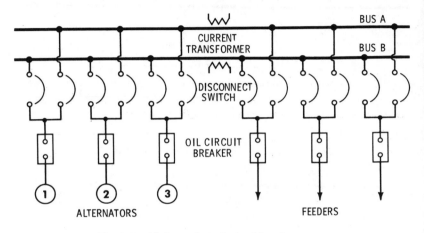

Fig. 9. Double-bus and single circuit-breaker system.

of a bus failure. It also permits continuance of service when maintenance or repair work is required on either bus. However, a service outage is unavoidable in case of a fault on the corresponding circuit breaker. Its principal advantage over the single-bus, single circuit-breaker system is that when a feeder trips out, it can be tested first on the other bus before returning it to normal service on the first main bus. This system is used in small generating stations.

Quite often, a tie-bus breaker is provided, as shown in Fig. 10. It facilitates line testing or quick transferring of power from one bus to the other. With the tie-bus breaker closed, thereby energizing both sets of bus bars, the transfer of a circuit carrying power from one to the other can be accomplished without danger of interruption to service by means of the disconnecting switches. Compare the tie-bus system in Fig. 10 with the spare-bus, spare circuit-breaker system shown in Fig.

426

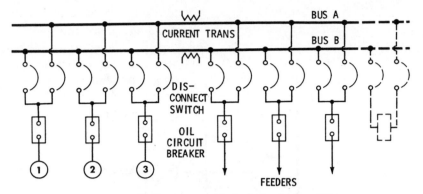

Fig. 10. Double-bus system with the tie bus in dotted line.

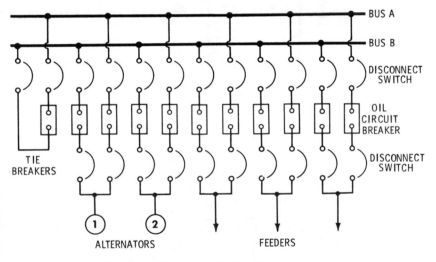

Fig. 11. Double-bus, double circuit-breaker system.

8. The tie-bus system is also confined chiefly to small generating stations.

Maximum flexibility is provided by the double-bus, double-breaker system, shown in Fig. 11. It has all the advantages of the double-bus, single-breaker system, with added assurance against shutdown of any particular circuit due to circuit-breaker faults. It is a comparatively expensive arrangement, used chiefly in large generating stations where continuity of service is of prime importance and the high cost is justi-

427

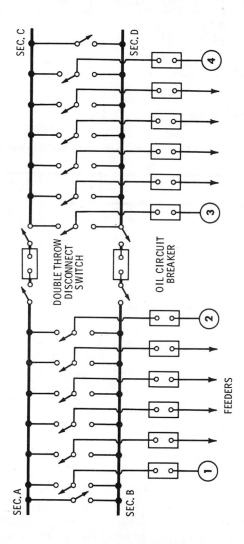

Fig. 12. Ring-bus system.

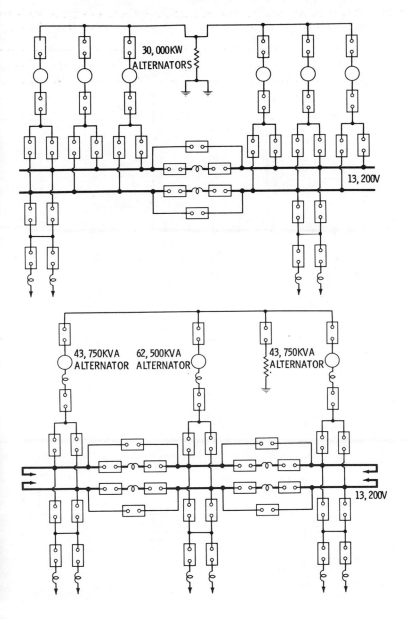

Fig. 13. Typical example of an "H" system.

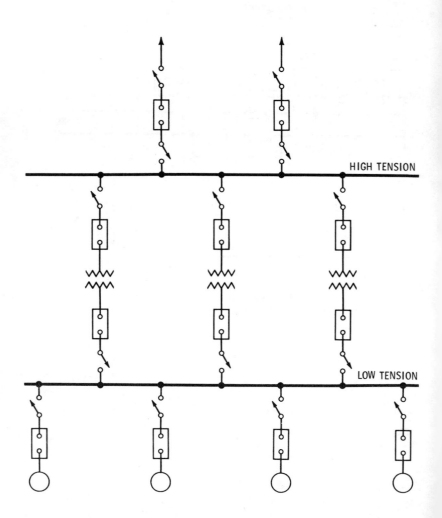

Fig. 14. System using single low- and single high-tension bus.

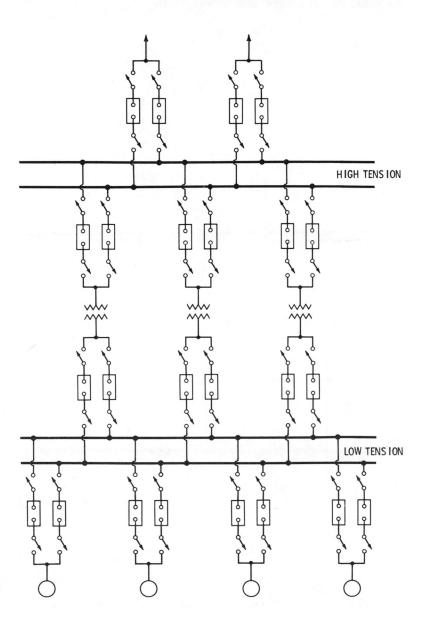

Fig. 15. Bus system using double buses and double circuit breakers.

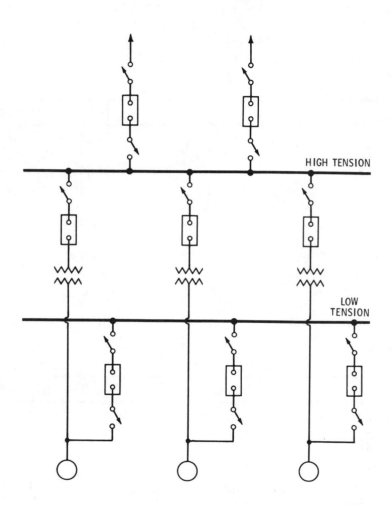

Fig. 16. Common arrangement where station is at a distance from the load center.

fied. Electricians in generating stations also work with other arrangements which differ slightly from the double-bus, double-breaker system, and the single-bus system.

The variations in connection systems have the purpose of providing the degree of flexibility which are warranted by local conditions, while minimizing the cost of the system. For example, Fig. 12 shows a modification of the double-bus, single-breaker system which is called a ring bus. The two buses are tied together by means of bus-tie circuit-breakers and disconnecting switches. In the H system, two feeder circuits are served from a pair of selector switches to either of two buses, as shown in the examples of Figs. 13A and 13B. This arrangement requires two breakers per feeder, and three breakers per generator.

For systems that distribute all or part of their power through step-up transformers, we will find more complex bus-bar connection arrangements. For example, Fig. 14 shows a service arrangement from a single low-tension bus to a single high-tension bus. High-tension (high-voltage) buses are constructed from copper, aluminum, or galvanized-steel tubing, bars, or other shapes, supported by insulators. Large buses are surrounded by fireproof barriers with considerable clearance space. Smaller buses may be mounted on a pipe or structural-steel framework located in an open space.

A step-up transformer arrangement using double buses and double breakers on both high- and low-tension circuits is shown in Fig. 15. If the station is located at some distance from the load center, particularly in the case of a hydroelectric station, a commonly-used bus arrangement is employed as shown in Fig. 16. Each alternator and step-up transformer is treated as a unit, and all power is transmitted over two or more lines. The load center (or load area) is not a generating station, but a substation which is fed from one or more generating stations.

Paralleling Generators

Auxiliary power for the station is obtained from a low-tension bus which may be connected to any one of the alternators. Normally, alternators are not paralleled on the low-tension bus. This arrangement is economical, but lacks flexibility. Let us briefly consider some of the factors that are involved in parallel operation of generators. The generator voltages must be equal, and their instantaneous polarities must

433

(A) Meter.

Fig. 17. Phase-

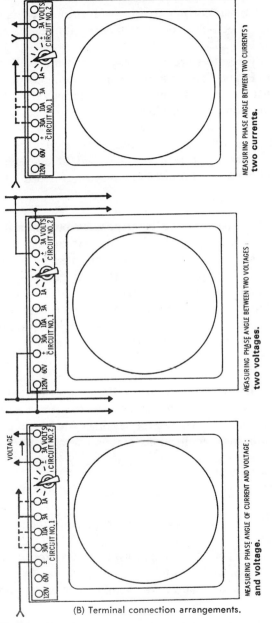

MEASURING PHASE ANGLE BETWEEN TWO CURRENTS : two currents.

MEASURING PHASE ANGLE BETWEEN TWO VOLTAGES : two voltages.

MEASURING PHASE ANGLE OF CURRENT AND VOLTAGE : and voltage.

(B) Terminal connection arrangements.

angle meter.

be the same. Generators connected in parallel must operate in phase with each other. Electricians describe this requirement as *phasing out* and *synchronizing*.

Three-phase generators are commonly used, with individual phases identified as A, B, and C. *Synchronizing* means bringing the generators into step so that their positive and negative alternation peaks occur at exactly the same time. If the alternators were connected in parallel when the phase difference is more than a very few degrees, heavy current surges would flow in the conductors between the alternators. Hence, phase-indicating instruments are used to determine when a pair of generators are synchronized and may be switched into a system for parallel operation.

A phase-angle meter is illustrated in Fig. 17A. The dial is calibrated in electrical degrees and corresponding power factor. Circuit No. 1, which is generally employed as the current circuit, has rated ranges of 1, 3, 10, and 30 amperes, supplemented by voltage ranges of 60 and 120 volts. Circuit No. 2, which is generally utilized as the voltage circuit, has switch-selected ranges of 15, 30, 60, 120, 240, and 480 volts, supplemented by a 3-ampere current range. In turn, the instrument can be used to measure the phase angle between a current and a voltage, or between two voltages, or between two currents.

With circuit No. 2 energized alone, the pointer slowly rotates on the 360° scale. Energization of circuit No. 1 stops the pointer rotation and indicates the phase angle between the sources, as well as whether circuit No. 1 is leading or lagging. An instrument of this type has the following applications:

1. Measurement of phase angles.
2. Connecting and checking directional and differential relays and their associated circuits.
3. Connecting and checking current transformers or voltage transformers.
4. Connecting and checking polyphase transformers, rectifier transformers, and so on.
5. Checking and connecting watthour meters, wattmeters, power-factor meters, synchroscopes, and their related circuits.

436

(A synchroscope is an instrument that indicates when one generator is brought exactly into phase with another generator for parallel operation).

6. Checking phase rotation (clockwise or counterclockwise) in polyphase systems.
7. Phasing buses for correct connection.
8. Use as a portable power-factor meter or synchroscope. Terminal connection arrangements are depicted in Fig. 17B.

Distribution

Each alternator in Fig. 16 can be used only as a unit on the corresponding transformer, and failure of either of the conductors between them will result in a shutdown of both. Furthermore, a failure of the high-tension bus will result in a complete shutdown of the plant until repairs are completed. To obtain somewhat greater flexibility, the method of high-tension connection shown in Fig. 18 is often used. It treats the transformer bank as part of the transmission line rather than as a unit with the alternator. A *transmission line* is defined as a line used for electric power transfer.

In Fig. 18, the transformer capacity (rating) is chosen with respect to the capacity of the transmission line. This arrangement shows three high-tension breakers per group. Often, to reduce cost, the two oil breakers A and B are replaced by three-pole air-break disconnect switches which may be operated electrically or manually. When this substitution is made, breaker C operates as both a line-breaker and as a transformer-breaker. The low-tension breaker D is often used to trip out the circuit, as the transformer would be considered as part of the line. The advantage of this arrangement is that, when operating on the low-tension side, the magnitude of voltage surges resulting from high-tension switching is minimized.

Some of the disadvantages of the arrangement shown in Fig. 18 are that it does not work well in a network system, nor is it economical when generating stations supply widely separated loads A network primary distribution system is an arrangement in which the primaries of the distribution transformers are connected to a common network supplied from the generating station. A primary distribution system is simply an arrangement for supplying the primaries of distribution transformers from a generating station. The term *distribution* includes

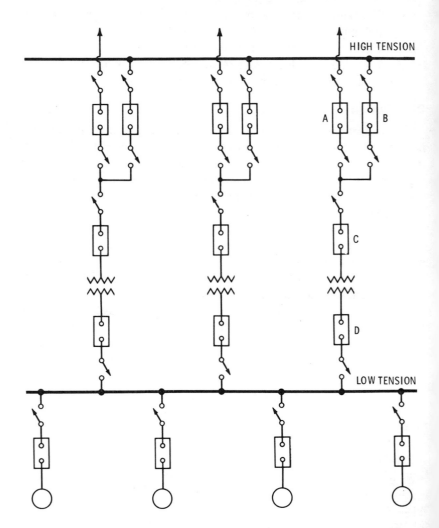

Fig. 18. System in which transformers are treated as a part of the transmission line.

all parts of an electric utility system.

The arrangement shown in Fig. 18 has feasibility in a station where power is to be transmitted over a number of lines to a single substation. In this example, the line and transformer banks are identical. If the line becomes lost due to a fault, the corresponding transformer cannot be used. Banking of distribution transformers means the tying together of secondary mains of adjacent transformers that are supplied by the same primary feeder. One circuit supplies all the transformers that have banked secondaries.

Fig. 19 shows an arrangement used in large steam generating plants where all power is fed into a high-tension network which is distributed over a considerable area. This station is called the *base load* plant. A base load is defined as the minimum load over a given period of time. With this arrangement, each alternator and step-up transformer is treated as a unit, with no switching devices between them. The high-tension bus is a straight double bus with a double breaker arrangement, which provides maximum flexibility. The power for station auxiliary is obtained from a high-tension step-down transformer bank.

Generating station auxiliaries are the accessory units of equipment necessary for plant operation, such as pumps, stokers, fans, etc. Essential auxiliaries are those which do not sustain service interruptions of more than one minute, such as boiler feed pumps, forced-draft fans, fuel feeders, etc. Nonessential auxiliaries may sustain service interruptions up to 3 minutes or more, such as air pumps, clinker grinders, coal crushers, and so on. Generating station auxiliary power is defined as the power that is required for operation of the auxiliaries.

Fig. 20 shows another station load plant in which all power is delivered to a high-tension bus. In this example, the transformer banks are of exceptionally large size, and the alternators are in two or three units, the steam end consisting of one high-pressure and one or two low-pressure turbines. A maximum degree of flexibility is provided, inasmuch as double buses are provided in both the high- and low-tension sections.

Figs. 21 through 24 show some of the more important and commonly used arrangements of main circuits in generating stations. Primary distribution mains are the conductors which feed from the center of distribution to direct primary loads or to transformers that

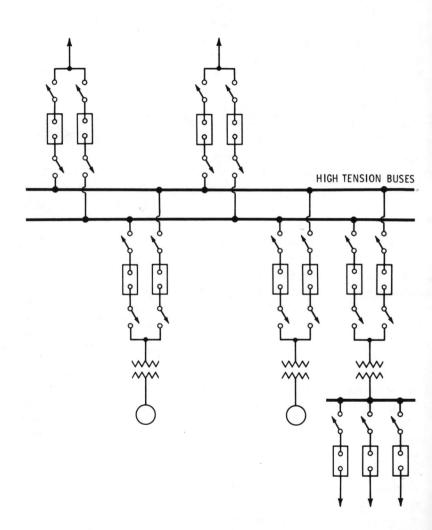

HIGH TENSION BUSES

Fig. 19. Bus system for large steam stations where power is fed into a H.T. network.

440

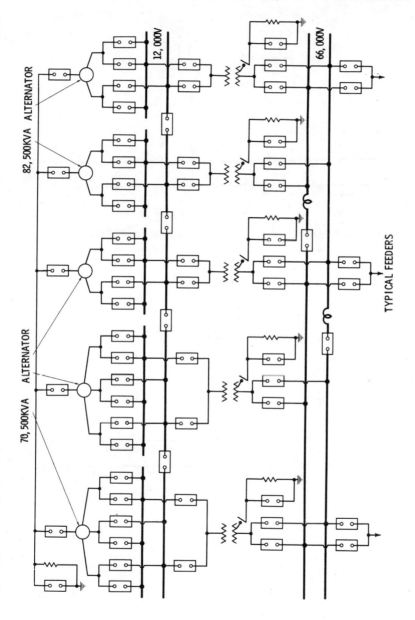

Fig. 20. Bus system where all power is delivered to a H.T. bus.

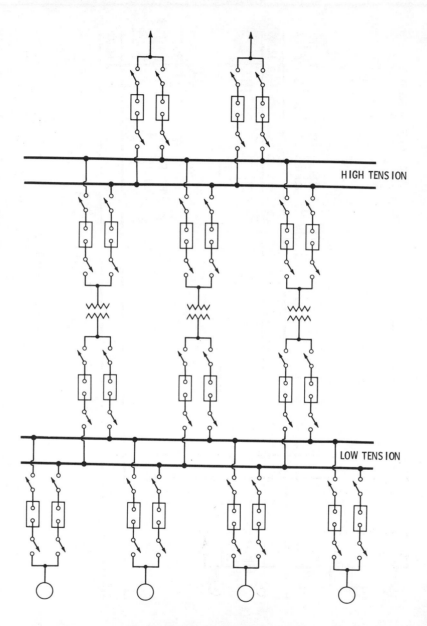

Fig. 21. Single sectionalized bus system.

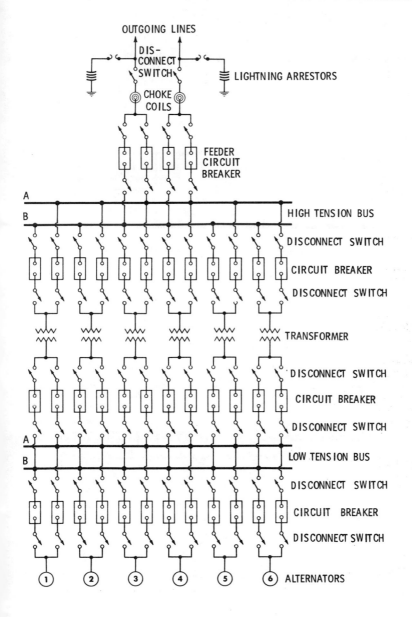

Fig. 22. Double-bus, double circuit-breaker system.

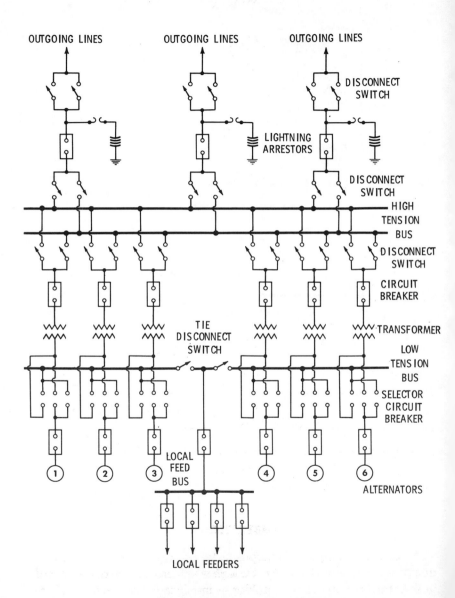

OUTGOING LINES   OUTGOING LINES   OUTGOING LINES

DISCONNECT SWITCH

LIGHTNING ARRESTORS

DISCONNECT SWITCH

HIGH TENSION BUS

DISCONNECT SWITCH

CIRCUIT BREAKER

TIE DISCONNECT SWITCH

TRANSFORMER

LOW TENSION BUS

SELECTOR CIRCUIT BREAKER

LOCAL FEED BUS

ALTERNATORS

LOCAL FEEDERS

Fig. 23. Single low-tension, double high-tension bus.

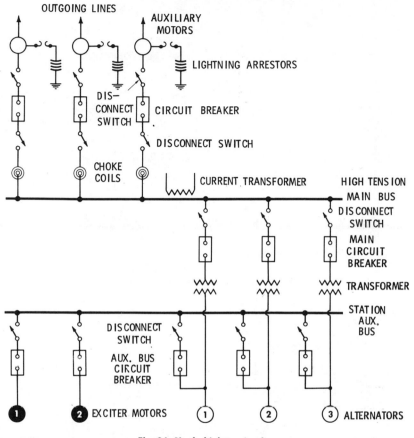

Fig. 24. Single high-tension bus.

feed secondary circuits. A distribution center is a point where equipment is located consisting generally of automatic overload protective devices connected to buses.

SUBSTATIONS

A substation consists of equipment for purposes other than generation or utilization of electricity, through which electric energy is passed in bulk for the purpose of switching or modification of electrical characteristics. A substation includes one or more buses, various circuit

445

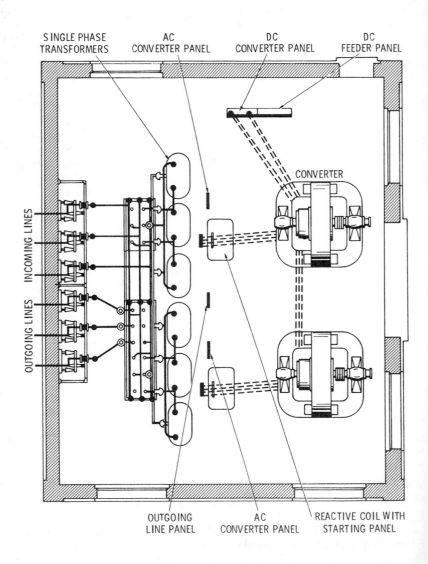

Fig. 25. Plans of a small substation.

breakers, and is the receiving point for one or more supply circuits. It may also sectionalize the transmission circuits that pass through it by means of circuit breakers.

Substations may be manually operated, semiautomatic, automatic, portable, or supervisory controlled. Fig. 25 shows the layout of a small substation. Substation transformers produce considerable heat, which must be dissipated by suitable means. Small transformers radiate heat from their shells; medium sizes have corrugated shells to increase the surface area. Small distribution substations may have no reserve transformer capacity, because transformer failures are rare and replacement can be made rapidly when necessary. A spare transformer is provided in a large substation.

Automatic Substations

In order to eliminate the uncertainty and expense of manual operation, unattended or automatic substations are often used. An automatic substation goes into operation under predetermined conditions by an automatic sequence. It automatically maintains the required characteristics of service. It goes out of operation by automatic sequence under other predetermined conditions and provides protection against usual operating emergencies. An automatic substation is usually started by a load demand on that part of the system within its particular district. This is accomplished by a voltage relay.

An automatic substation is stopped by the operation of an undervoltage relay when the load diminishes to an uneconomical point. Starting and stopping may also be accomplished by a remote-control system, or by a time switch. The sequence of the various operations is determined by a motor-driven master switch. It makes and breaks circuits and actuates contactors and relays which act directly on the machine circuits. This ensures correct sequence of operation and also eliminates a large number of interlocks. An *interlock* is a device actuated by operation of another device, with which it is directly associated, to govern succeeding operations. Interlocks may be either electrical or mechanical.

Semiautomatic Substations

A *semiautomatic substation* is started manually and runs until shut down, according to some schedule, by one of a number of different

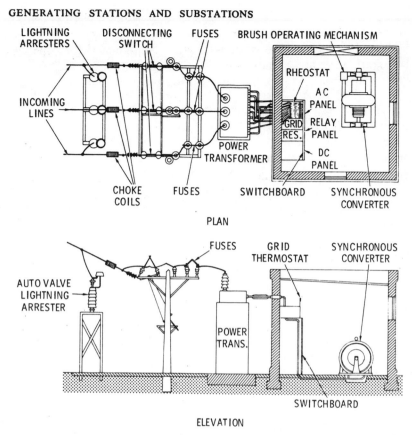

Fig. 26. Plan and elevation of typical outdoor semiautomatic substation.

methods. These methods include a time switch, momentary interruption of the AC supply, or by an attendant. Fig. 26 shows the layout of a semiautomatic substation. Since a semiautomatic substation is attended only during the starting and possibly during the shutting-down period, it must be equipped with all protective devices such as are included with fully automatic equipment. These devices prevent open-phase running, excessive temperature of machine or transformer windings, overheated bearings, operation with open shunt-field winding, and armature overspeed. Complete automatic operation of the DC equipment is essential.

To meet emergency conditions, it is essential that means be provided for quickly opening all feeders. In many installations, it is also desirable to shut down and lock out certain automatic substations during light-load periods. There is, accordingly, a requirement for supervising unattended automatic substations from a central point or dispatcher's station. Automatic supervisory equipment provides the dispatcher with a means of selectively controlling devices in the substations and automatically gives him a visual indication of the substation apparatus by means of indicating lamps located in cabinets at his office.

The dispatcher's office equipment consists of control keys, indicating lamps, and necessary supervisory devices for receiving control impulses which indicate the positions of the various supervised units. The substation equipment consists of the supervisory devices which transmit

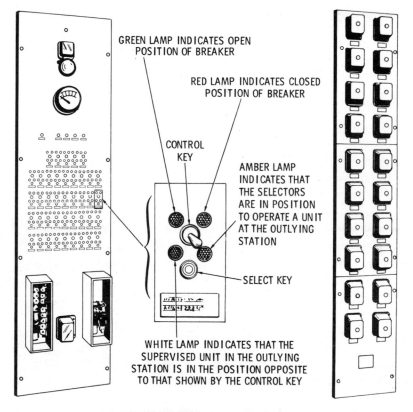

GREEN LAMP INDICATES OPEN POSITION OF BREAKER

RED LAMP INDICATES CLOSED POSITION OF BREAKER

CONTROL KEY

AMBER LAMP INDICATES THAT THE SELECTORS ARE IN POSITION TO OPERATE A UNIT AT THE OUTLYING STATION

SELECT KEY

WHITE LAMP INDICATES THAT THE SUPERVISED UNIT IN THE OUTLYING STATION IS IN THE POSITION OPPOSITE TO THAT SHOWN BY THE CONTROL KEY

Fig. 27. Dispatchers control panel.

449

the control impulses to the auxiliary control relays and send back indication impulses to the dispatcher's office. Typical supervisory equipment is illustrated in Fig. 27.

At the dispatcher's office, each supervised unit has a key and lamp combination consisting of a standard two-position turn key for control, a red light for indicating the closed position, and a white light for indicating an automatic operation of the corresponding breaker unit. Each combination has a two-position push-and-pull selecting key to stop the selectors at a point corresponding to the unit it is desired to control, and an associated amber lamp for indicating when the selectors are connected to that particular unit.

The dispatcher controls the supervised unit in the substation over the control circuit, and the indications from the supervised units are returned to the dispatcher over the indication circuit. The equipment at the substation is made to operate in synchronism with the equipment in the dispatcher's office by means of current impulses sent over the synchronizing circuit.

SUMMARY

A generating station is a plant where electric energy is produced. A substation is an assembly of equipment which electricity passes through for the purpose of switching or modifying its characteristics. A substation is usually a receiving point for more than one supply circuit.

A generating station deals with the installation of generating units and mechanical and auxiliary electrical equipment required to utilize mechanical energy, changing the mechanical energy into electrical energy. This energy is then delivered to the associated systems.

There are four classifications of generating stations; gas-powered, oil-powered, water-powered, and steam generating. The electrical features are quite similar in all four classes of stations. The turbines, alternators, exciters, and controlling switchboards are all housed in one large room of the hydroelectric generating station.

The main electrical connections in a generating station depends on the size and type of generator, the feeder arrangement, and function of the generating station in the system. The simplest connection arrangement is called the single-bus system, which is used in small stations where interrupted service is permissible.

To obtain greater flexibility and to reduce the possibility of service interruptions, two buses with additional disconnect switches or manually-operated oil circuit breakers may be used in small generating stations. One bus is called the main and the other is called the spare bus. Another connection arrangement that employs two buses is called the double-bus, single breaker system. It practically eliminates the possibility of a prolonged shutdown in case of a bus failure.

Substations may be manually operated, semiautomatic, automatic, portable, or supervisory controlled. In order to eliminate the uncertainty and expense of manual operation, unattended or automatic substations are often used. An automatic substation goes into operation under predetermined conditions by an automatic sequence.

Semiautomatic substations are started manually and run until shutdown by one of a number of different methods. These methods include a time switch, momentary interruption of the AC supply, or by an attendant's action. Since a semiautomatic station is attended only during the starting and during shutdown periods, it must be equipped with all protective devices such as those in fully automatic substations.

TEST QUESTIONS

1. Define a generating station.
2. What is meant by an automatic station?
3. Explain the purpose of a prime mover.
4. How does an exciter operate?
5. Describe a single-bus system for a generating station.
6. When does a circuit breaker open its circuit?
7. Discuss the operation of an instrument transformer.
8. What are the chief features of a spare-bus, spare circuit-breaker system?
9. Describe a double-bus, single circuit-breaker system.
10. Define a tie-bus breaker.
11. Explain the features of a double-bus, double-breaker system.

12. Discuss a ring-bus system.
13. What is meant by an "H" system?
14. How are transformers banked?
15. Define a base load.
16. Describe a single sectionalized bus system.
17. Explain the chief features of a double-bus, double circuit-breaker system.
18. What is an advantage of the single low-tension, double high-tension, single circuit-breaker arrangement?
19. Explain the arrangement of a single high-tension bus system.
20. Define an electric power substation.
21. What are the chief features of an automatic substation?
22. Explain what is meant by a semiautomatic substation.
23. State several functions of protective devices used in semiautomatic substations.
24. Describe the purposes served by automatic supervisory equipment.
25. How does a dispatcher determine the position of supervised units?

Glossary

Access Fitting—A fitting that permits access to conductors in concealed or enclosed wiring, elsewhere than at an outlet.

Active Electrical Network—A network that contains one or more sources of electrical energy.

Admittance—The reciprocal of impedance.

Air-Blast Transformer — A transformer cooled by forced circulation of air through its core and coils.

Air Circuit Breaker—A circuit breaker in which the interruption occurs in air.

Air Switch—A switch in which the interruption of the circuit occurs in air.

Alive—Electrically connected to a source of emf, or electrically charged with a potential different from that of the earth. Also: Practical synonym for "current-carrying."

Alternating Current—A periodic current, the average value of which over a period is zero.

Alternator — (Synchronous Generator); a synchronous alternating-current machine which changes mechanical power into electrical power.

Ambient Temperature — The temperature of a surrounding cooling medium, such as gas or liquid, which comes into contact with the heated parts of an apparatus.

Ammeter—An instrument for measuring electric current.

Ampere—A charge flow of one coulomb per second.

Annunciator—An electromagnetically operated signaling apparatus which indicates whether a current is flowing or has flowed in one or more circuits.

Apparent Power—In a single-phase, two-wire circuit, the product of the effective current in one conductor multiplied by the effective voltage between the two points of entry.

Appliance — Current-consuming equipment, fixed or portable, such as heating or motor-operated equipment.

Arcing Contacts—Contacts on which an arc is drawn after the main contacts of a switch or circuit-breaker have parted.

Arcing Time of Fuse—The time elapsing from the severance of the fuse link to the final interruption of the circuit under specified conditions.

Arc-over of Insulator—A discharge of power current in the form of an arc, following a surface discharge over an insulator.

Armor Clamp—A fitting for gripping the armor of a cable at the point where the armor terminates, or where the cable enters a junction box or other apparatus.

Armored Cable—A cable provided with a wrapping of metal, usually steel wires, primarily for the purpose of mechanical protection.

Arrester, Lightning—A device which reduces the voltage of a surge applied to its terminals, and restores itself to its original operating condition.

Autotransformer — A transformer in which part of the winding is common to both the primary and secondary circuits.

Back Connected Switch—A switch in which the current-carrying conductors are connected to studs in back of the mounting base.

Bank—An assemblage of fixed contacts in a rigid unit over which wipers or brushes may move and make connection with the contacts.

Bank, Duct—An arrangement of conduit which provides one or more continuous ducts between two points.

Benchboard—A switchboard with a horizontal section for control switches, indicating lamps, and instrument switches; may also have a vertical instrument section.

Bidirectional Current—A current which has both positive and negative values.

Bond, Cable—An electric connection across a joint in the armor or lead sheath of a cable, or between the armor or sheath to ground, or between the armor or sheath of adjacent cables.

Box, Conduit—A metal box adapted for connection to conduit for installation of wiring, making connections, or mounting devices.

Box, Junction—An enclosed distribution panel for connection or branching of one or more electric circuits without making permanent splices.

Box, Junction (Interior Wiring)—A metal box with blank cover for joining runs of conduit, electrical metallic tubing, wireway or raceway, and providing space for connection and branching of enclosed conductors.

Box, Pull—A metal box with a blank cover which is used in a run of conduit, etc., to facilitate pulling in the conductors; it may also be installed at the end of one or more conduit runs for distribution of the conductors.

Branch Circuit—That portion of a wiring system extending beyond the final automatic overload protective device.

Branch Circuit, Appliance—A circuit supplying energy either to permanently wired appliances or to attachment-plug receptacles such as appliance or convenience outlets, and having no permanently connected lighting fixtures.

Branch Circuit Distribution Center—A distribution circuit at which branch circuits are supplied.

Branch Circuit, Lighting—A circuit supplying energy to lighting outlets only.

Branch Conductor—A conductor that branches off at an angle from a continuous run of conductor.

Branch Joint — A multiple joint for connection of a branch conductor or cable to a main conductor or cable, wherein the latter continues beyond the branch.

Break—The break of a circuit-opening device is the minimum distance between the stationary and movable contacts when the device is in its open position.

Breakdown — Also termed "puncture", denoting a disruptive discharge through insulation.

Breaker, Line—A device that combines the functions of a contactor and a circuit-breaker.

Buried Cable—A cable installed under the surface of the soil in such manner that it cannot be removed without digging up the soil.

Bus—A conductor or group of conductors which serves as a common connection for three or more circuits in a switchgear assembly.

Bushing—Also termed "insulating bushing"; a lining for a hole for insulation and/or protection from abrasion of one or more conductors passing through it.

Cabinet — An enclosure for either surface or flush mounting, provided with a frame, mat, or trim.

Cable Fault—A partial or total local failure in the insulation or continuity of the conductor.

Cable Joint—Also termed a "splice"; a connection between two or more individual lengths of cables, with their conductors individually connected, and with protecting sheaths over the joint.

Cable, Service—Service conductors arranged in the form of a cable.

Cable Sheath—The protective covering, such as lead, applied over a cable.

Charge, Electric—An inequality of positive and negative electricity in or on a body. The charge stored in a capacitor (condenser) corresponds to a deficiency of free electrons on the positive plate, and to an excess of free electrons on the negative plate.

Choke Coil—A low-resistance coil with sufficient inductance to substantially impede AC or transient currents.

Circuit, Electric—A conducting path through which electric charges may flow. A DC circuit is a closed path for charge flow; an AC circuit is not necessarily closed, and may conduct in part by means of an electric field (displacement current).

Circuit, Earth (Ground) Return—An electric circuit in which the ground serves to complete a path for charge flow.

Circuit, Magnetic—A closed path for establishment of magnetic flux (magnetic field) which has the direction of the magnetic induction at every point.

Cleat—An assembly of a pair of insulating material members with grooves for holding one or more conductors at a definite distance from the mounting surface.

Clip, Fuse — Contacts on a fuse support for connecting a fuse holder into a circuit.

Closed-Circuit Voltage — The terminal voltage of a source of electricity under a specified current demand.

Closed Electric Circuit — A continuous path or paths providing for charge flow. In an AC closed circuit, charge flow may be changed into displacement current "through" a capacitor (condenser).

Coercive Force — The magnetizing force at which the magnetic induction is zero at a point on the hysteresis loop of a magnetic substance.

Coil—A conductor arrangement (basically a helix or spiral) that concentrates the magnetic field produced by electric charge flow.

Composite Conductor—A conductor consisting of two or more strands of different metals, operated in parallel.

Concealed—To be made inaccessible by the structure or finish of a building; also, wires run in a concealed raceway.

Condenser—Also termed "capacitor"; a device that stores electric charge by means of an electric field.

Conductance—A measure of permissiveness to charge flow; the reciprocal of resistance.

Conductor—A substance that has free electrons or other charge carriers which permit charge flow when an emf is applied across the substance.

Conduit—A structure containing one or more ducts; commonly formed from iron pipe or electrical metallic tubing.

Conduit Fittings—Accessories used to complete a conduit system, such as boxes, bushings, and access fittings.

Conduit, Flexible Metal — A flexible raceway of circular form for enclosing wires or cables; usually made of steel wound helically and with interlocking edges, and with a weather-resistant coating.

Conduit, Rigid Steel—A raceway made of mild steel pipe with a weather-resistant coating.

Conduit Run—A duct bank; an arrangement of conduit with a continuous duct between two points in an electrical installation.

Contactor—An electric power switch, not operated manually, and designed for frequent operation.

Contacts — Conducting parts which employ a junction that is opened or closed to interrupt or complete a circuit.

Control Relay—A relay used to initiate or permit a predetermined operation in a control circuit.

Coulomb—An electric charge of 6.28×10^{18} electrons. One coulomb is transferred when a current of 1 ampere continues past a point for one second.

Counter EMF—(CEMF); the effective emf within a system which opposes current in a specified direction.

Current—The rate of charge flow. A current of one ampere is equal to a flow rate of one coulomb per second.

Cycle—The complete series of values which occur during one period of a periodic quantity. The

457

unit of frequency, the hertz, is equal to one cycle per second.

Dead—Functionally conducting parts of an electrical system that have no potential difference or charge (voltage of zero with respect to ground).

Degree, Electrical—An angle equal to 1/360 of the angle between consecutive field poles of like polarity in an electrical machine.

Diagram, Connection — A drawing showing the connections and interrelations of devices employed in an electrical circuit.

Dielectric—A medium or substance in which a potential difference establishes an electric field which is subsequently recoverable as electric energy.

Direct Current—A unidirectional current with a constant value. "Constant value" is defined in practice as a value that has negligible variaton.

Direct EMF—Also termed "direct voltage"; an emf that does not change in polarity and has a constant value (one of negligible variation).

Discharge—An energy conversion involving electrical energy. Examples: discharge of a storage battery; discharge of a capacitor; lightning discharge of a thundercloud.

Displacement Current — The apparent flow of charge "through" a dielectric such as in a capacitor; represented by buildup and/or decay of an electric field.

Disruptive Discharge—A rapid and large current increase through an insulator due to insulation failure.

Distribution Center—A point of installation for automatic overload protective devices connected to buses where an electrical supply is subdivided into feeders and/or branch circuits.

Divider, Voltage—A tapped resistor or series arrangement of resistors, sometimes with movable contacts, providing a desired IR drop. (A voltage divider is not continuously and manually variable as in a potentiometer).

Drop, Voltage—An IR voltage between two specified points in an electric circuit.

Duct—A single enclosed runway for conductors or cables.

Effective Value—The effective value of a sine-wave AC current or voltage is equal to 0.707 of peak. Also called the root-mean-square (rms) value, it produces the same I^2R power as an equal DC value.

458

Efficiency—The ratio of output power to input power, usually expressed as a percentage.

Electrical Units—In the practical system, electrical units comprise the volt, the ampere, the ohm, the watt, the watt-hour, the coulomb, the mho, the henry, the farad, and the joule.

Electricity—A physical entity associated with the atomic structure of matter which occurs in polar forms (positive and negative) and which are separable by expenditure of energy.

Electrode—A conducting substance through which electric current enters or leaves in devices that provide electrical control or energy conversion.

Electrolyte—A substance that provides electrical conduction when dissolved (usually in water.)

Electrolytic Conductor—Flow of electric charges to and from electrodes in an electrolytic solution.

Electromagnetic Induction—A process of generation of emf by movement of magnetic flux which cuts an electrical conductor.

Electromotive Force—(EMF); an energy-charge relation that results in electric pressure which produces or tends to produce charge flow.

Electron—The subatomic unit of negative electricity; it is a charge of 1.6×10^{-19} coulomb.

Electronics—The science treating of charge flow in vacuum, gases, and crystal lattices.

Electroplating—The electrical deposition of metallic ions as neutral atoms on an electrode immersed in an electrolyte.

Electrostatics — A branch of electrical science dealing with the laws of electricity at rest.

Energy—The amount of physical work which a system is capable of doing. Electrical energy is measured in watt-seconds, or the product of power and time.

Entrance, Duct—An opening of a duct at a distributor box or other accessible location.

Equipment, Service—A circuit-breaker or switches and fuses with their accessories, installed near the point of entry of service conductors to a building.

Exciter—An auxiliary generator for supplying electrical energy to the field of another electrical machine.

Farad—A unit of capacitance that is defined by the production of one volt across the capacitor terminals when a charge of one coulomb is stored.

459

Fault Current—An abnormal current flowing between conductors or from a conductor to ground due to an insulation defect, arcover, or incorrect connection.

Feeder—A conductor or a group of conductors for connection of generating stations, substations, generating and substations, or a substation and a feeding point.

Ferromagnetic Substance — A substance that has a permeability considerably greater than that of air; a ferromagnetic substance has a permeability that changes with the value of applied magnetizing force.

Filament—A wire or ribbon of conducting (resistive) material which develops light and heat energy due to electric charge flow; light radiation is also accompanied by electron emission.

Fixture Stud — A fitting for mounting a lighting fixture in an outlet box, and which is secured to the box.

Flashover—A disruptive electrical discharge around or over (but not through) an insulator.

Fluorescence — An electrical discharge process involving radiant energy transferred by phosphors into radiant energy that provides increased luminosity.

Flux — Electrical field energy distributed in space, in a magnetic substance, or in a dielectric. Flux is commonly represented diagramatically by means of flux lines denoting magnetic or electric forces.

Force—An elementary physical cause capable of modifying the motion of a mass.

Frequency — The number of periods occurring in unit time of a periodic process such as in the flow of electric charge.

Frequency Meter—An instrument that measures the frequency of an alternating current.

Fuse — A protective device with a fusible element that opens the circuit by melting when subjected to excessive current.

Fuse Cutout — An assembly consisting of a fuse support and holder, which may also include a fuse link.

Fuse Element — Also termed "fuse link"; the current-carrying part of a fuse which opens the circuit when subjected to excessive current.

Fuse Holder — A supporting device for a fuse that provides terminal connections.

Galvanometer—An instrument for indicating or measuring comparatively small electric currents. A galvanometer usually has zero-center indication.

Gap— (Spark Gap); a high-voltage device with electrodes between which a disruptive discharge of electricity may pass, usually through air. A sphere gap has spherical electrodes; a needle gap has sharply pointed electrodes; a rod gap has rods with flat ends.

Ground—Also termed "earth"; a conductor connected between a circuit and the soil; a chassis-ground is not necessary at ground potential, but is taken as a zero-volt reference point. An accidental ground occurs due to cable insulation faults, an insulator defect, etc.

Grounding Electrode—A conductor buried in the earth, for connection to a circuit. The buried conductor is usually a cold-water pipe, to which connection is made with a ground clamp.

Ground Lug—A lug for convenient connection of a grounding conductor to a grounding electrode or device to be grounded.

Ground Outlet—An outlet provided with a polarized receptacle with a grounded contact for connection of a grounding conductor.

Ground Switch—A switch for connection or disconnection of a grounding conductor.

Guy—A wire or other mechanical member having one end secured and the other end fastened to a pole or structural part maintained under tension.

Hanger—Also termed "cable rack"; a device usually secured to a wall to provide support for cables.

Heat Coil—A protective device for opening and/or grounding a circuit by switching action when a fusible element melts due to excessive current.

Heater—In the strict sense, a heating element for raising the temperature of an indirectly heated cathode in a vacuum or gas tube. Also applied to appliances such as space heaters and radiant heaters.

Henry—The unit of inductance; it permits current increase at the rate of 1 ampere per second when 1 volt is applied across the inductor terminals.

Hickey—A fitting for mounting a lighting fixture in an outlet box. Also, a device used with a pipe handle for bending conduit.

Horn Gap—A form of switch provided with arcing horns for automatically increasing the length of the arc and thereby extinguishing the arc.

461

Hydrometer — An instrument for indicating the state of charge in a storage battery.

Hysteresis — The magnetic property of a substance which results from residual magnetism.

Hysteresis Loop—A graph that shows the relation between magnetizing force and flux density for a cyclically magnetized substance.

Hysteresis Loss—The heat loss in a magnetic substance due to application of a cyclic magnetizing force to a magnetic substance.

Impedance—Opposition to AC current by a combination of resistance and reactance; impedance is measured in ohms.

Impedances, Conjugate — A pair of impedances that have the same resistance values, and that have equal and opposite reactance values.

Impulse—An electric surge of unidirectional polarity.

Indoor Transformer—A transformer that must be protected from the weather.

Induced Current—A current that results in a closed conductor due to cutting of lines of magnetic force.

Inductance — An electrical property of a resistanceless conductor which may have a coil form, and which exhibits induc-

tive reactance to an AC current. All practical inductors have at least a slight amount of resistance, also.

Inductor—A device such as a coil with or without a magnetic core which develops inductance, as distinguished from the inductance of a straight wire.

Instantaneous Power—The product of an instantaneous voltage by the associated instantaneous current.

Instrument—An electrical device for measurement of a quantity under observation, or for presenting a characteristic of the quantity.

Interconnection, System — A connection of two or more power systems.

Interconnection Tie—A feeder that interconnects a pair of electric supply systems.

Interlock—An electrical device depending on its operation from another device, for controlling subsequent operations.

Internal Resistance — The effective resistance connected in series with a source of emf due to resistance of the electrolyte, winding resistance, etc.

Ion—A charged atom, or a radical. For example, a hydrogen atom that has lost an electron becomes a hydrogen ion; sulphuric

462

acid produces H⁺ and SO⁻₄ ions in water solution.

IR Drop—A potential difference produced by charge flow through a resistance.

Isolating Switch—An auxiliary switch for isolating an electric circuit from its source of power; it is operated only after the circuit has been opened by other means.

Joule—A unit of electrical energy; also called a watt-second; the transfer of one watt for one second.

Joule's Law — The rate at which electrical energy is changed into heat energy is proportional to the square of the current.

Jumper — A short length of conductor for making a connection between terminals, around a break in a circuit, or around an electrical instrument.

Junction—A point in a parallel or series-parallel circuit where current branches off into two or more paths.

Junction Box — An enclosed distribution panel for the connection or branching of one or more electrical circuits, not using permanent splices. In the case of interior wiring, a junction box consists of a metal box with a blank cover; it is inserted in a run of conduit, raceway, or tubing.

Kirchhoff's Law—The voltage law states that the algebraic sum of the drops around a closed circuit is equal to zero. The current law states that the algebraic sum of the currents at a junction is equal to zero.

Knockout—A scored portion in the wall of a box or cabinet which can be removed easily by striking with a hammer; a circular hole is provided thereby for accommodation of conduit or cable.

KVA — Kilovolt-amperes; the product of volts and amperes divided by 1,000.

Lag—Denotes that a given sine wave passes through its peak at a later time than a reference sine wave.

Lampholder — Also termed "socket" or "lamp receptacle"; a device for mechanical support of and electrical connection to a lamp.

Lay—The lay of a helical element of a cable is equal to the axial length of a turn.

Lead — Denotes that a given sine wave passes through its peak at an earlier time than a reference sine wave.

Leakage, Surface—Passage of current over the boundary surfaces of an insulator as distinguished from passage of current through its bulk.

Leg of a Circuit—One of the conductors in a supply circuit between which the maximum supply voltage is maintained.

Lenz' Law—States that an induced current in a conductor is in a direction such that the applied mechanical force is opposed.

Limit Switch—A device that automatically cuts the power off at or near the limit of travel of a mechanical member.

Load—The load on an AC machine or apparatus is equal to the product of the rms voltage across its terminals and the rms current demand.

Locking Relay—A relay which operates to make some other device inoperative under certain conditions.

Loom—See *Tubing, Flexible*.

Luminosity—Relative quantity of light.

Magnet—A magnet is a body which is the source of a magnetic field.

Magnetic Field—A magnetic field is the space containing distributed energy in the vicinity of a magnet, and in which magnetic forces are apparent.

Magnetizing Force—Number of ampere-turns in a transformer primary per unit length of core.

Magnetomotive Force—Number of ampere-turns in a transformer primary.

Mass—Quantity of matter; the physical property which determines the acceleration of a body as the result of an applied force.

Matter—Matter is a physical entity that exhibits mass.

Meter—A unit of length equal to 39.37 inches; an electrical instrument for measurement of voltage, current, power, energy, phase angle, synchronism, resistance, reactance, impedance, inductance, capacitance, etc.

Mho—The unit of conductance defined as the reciprocal of the ohm.

Mounting, Circuit-Breaker — Supporting structure for a circuit-breaker.

Multiple Feeder — Two or more feeders connected in parallel.

Multiple Joint — A joint for connecting a branch conductor or cable to a main conductor or cable, to provide a branch circuit.

Multiplier, Instrument—A series resistor connected to a meter mechanism for the purpose of providing a higher votlage-indicating range.

Mutual Inductance — An inductance which is common to the primary and secondary of a transformer, resulting from primary magnetic flux that cuts the secondary winding.

Negative—A value less than zero; an electric polarity sign indicating an excess of electrons at one point with respect to another point; a current sign indicating charge flow away from a junction.

Network—A system of interconnected paths for charge flow.

Network, Active—A network that contains one or more source of electrical energy.

Network, Passive—A network that does not contain a source of electrical energy.

No-Load Current — The current demand of a transformer primary, when no current demand is made on the secondary.

Normally Closed—Denotes the automatic closure of contacts in a relay when deenergized (not applicable to a latching relay).

Normally Open—Denotes the automatic opening of contacts in a relay when deenergized. (Not applicable to a latching relay.)

Ohm—The unit of resistance; a resistance of one ohm sustains a current of one ampere when one volt is applied across the resistance.

Ohmmeter—An instrument for measuring resistance values.

Ohm's Law—States that current is directly proportional to applied voltage, and inversely proportional to resistance, reactance, or impedance.

Open-Circuit Voltage — The terminal voltage of a source under conditions of no current demand. The open-circuit voltage has a value equal to the emf of the source.

Open-Wire Circuit—A circuit constructed from conductors that are separately supported on insulators.

Oscilloscope — An instrument for displaying the waveforms of AC voltages.

Outdoor Transformer — A transformer with weatherproof construction.

Outlet—A point in a wiring system from which current is taken for supply of fixtures, lamps, heaters, etc.

Outlet, Lighting — An outlet used for direct connection of a lampholder, lighting fixture, or a cord that supplies a lampholder.

Outlet, Receptacle—An outlet used with one or more receptacles which are not of the screw-shell type.

Overload Protection — Interruption or reduction of current under conditions of excessive demand, provided by a protective device.

Ozone—A compound consisting of three atoms of oxygen, produced by the action of electric sparks, or specialized electrical devices.

Peak Current—The maximum value (crest value) of an alternating current.

Peak Voltage—The maximum value (crest value) of an alternating voltage.

Peak-to-Peak Value — The value of an AC waveform from its positive peak to its negative peak. In the case of a sine wave, the peak-to-peak value is double the peak value.

Pendant—A fitting suspended from overhead by a flexible cord which may also provide electrical connection to the fitting.

Pendant, Rise-and-Fall — A pendant that can be adjusted in height by means of a cord adjuster.

Period—The time required for an AC waveform to complete one cycle.

Permanent Magnet — A magnetized substance that has substantial retentivity.

Permeability — The ratio of magnetic flux density to magnetizing force.

Phase—The time of occurrence of the peak value of an AC waveform with respect to the time of occurrence of the peak value of a reference waveform. Phase is usually stated as the fractional part of a period.

Phase Angle—An angular expression of phase difference; it is commonly expressed in degrees, and is equal to the phase multiplied by 360°.

Plug—A device which is inserted into a receptacle for connection of a cord to the conductor terminations in the receptacle.

Polarity—An electrical characteristic of emf which determines the direction in which current tends to flow.

Polarization (Battery)—Polarization is caused by development of gas at the battery electrodes during current demand, and has the effect of increasing the internal resistance of the battery.

Pole—The pole of a magnet is an area at which its flux lines tend to converge or diverge.

Positive—A value greater than zero; an electric polarity sign denoting a deficiency of electrons at one point with respect to another point; a current sign indicating charge flow toward a junction.

Potential Difference — A potential difference of one volt is produced when one unit of work is done in separating unit charges through unit distance.

Potentiometer—A resistor with a continuously variable contact arm; electrical connections are made to both ends of the resistor and to the arm.

Power — The rate of doing work, or the rate of converting energy. When one volt is applied to a load and the current demand is one ampere, the rate of energy conversion (power) is one watt.

Power, Real—Real power is developed by circuit resistance, or effective resistance.

Primary Battery — A battery that cannot be recharged after its chemical energy has been depleted.

Primary Winding—The input winding of a transformer.

Proton—The subatomic unit of positive charge; a proton has a charge which is equal and opposite to that of an electron.

Pull Box—A metal box with a blank cover for insertion into a conduit run, raceway, or metallic tubing, which facilitates the drawing of conductors.

Pulsating Current — A direct current which does not have a steady value.

Puncture — A disruptive electrical discharge through insulation.

Quick-Break—A switch or circuit-breaker which has a high contact-opening speed.

Quick-Make—A switch or circuit breaker which has a high contact-closing speed.

Raceway—A channel for holding wires or cables; constructed from metal, wood, or plastics, rigid metal conduit, electrical metal tubing, cast-in-place, underfloor, surface metal, surface wooden types, wireways, busways, and auxiliary gutters.

Rack, Cable — A device secured to the wall to provide support for a cable raceway.

Rating—The rating of a device, apparatus, or machine states the limit or limits of its operating characteristics. Ratings are commonly stated in volts, amperes, watts, ohms, degrees, horsepower, etc.

Reactance — Reactance is an opposition to AC current based on the reaction of energy storage, either as a magnetic field or as an electric field. No real power is dissipated by a reactance. Reactance is measured in ohms.

467

Reactor — An inductor or a capacitor. Reactors serve as current-limiting devices such as in motor starters, for phase-shifting applications as in capacitor start motors, and for power-factor correction in factories or shops.

Receptacle—Also termed "convenience outlet"; a contacting device installed at an outlet for connection externally by means of a plug and flexible cord.

Rectifier—A device which has a high resistance in one direction, and a low resistance in the other direction.

Regulation — Denotes the extent to which the terminal voltage of a battery, generator, or other source decreases under current demand. Commonly expressed as the ratio of the difference of the no-load voltage and the load voltage to the no-load voltage under rated current demand; usually expressed as a percentage.

Relay—A device operated by a change in voltage or current in a circuit, which actuates other devices in the same circuit or in another circuit.

Reluctance—An opposition to the establishment of magnetic flux lines when a magnetizing force is applied; usually measured in rels.

Remanence—The flux density that remains in a magnetic sub-stance after an applied magneto-motive force has been removed.

Resistance—A physical property which opposes current and dissipates real power in the form of heat. Resistance is measured in ohms.

Resistor — A resistive component; may be of the wirewound, carbon - composition, thyrite, or other type of design.

Rheostat—A variable resistive device consisting of a resistance element and a continuously adjustable contact arm.

Rosette—A porcelain or other enclosure with terminals for connecting a flexible cord and pendant to the permanent wiring.

Safety Outlet — Also termed "ground outlet"; an outlet with a polarized receptacle for equipment grounding.

Secondary Battery—A battery which can be recharged after its chemical energy is depleted.

Sequence Switch—A remotely controlled power-operated switching device.

Series Circuit—A circuit that provides a complete path for current and has its components connected end-to-end.

Service—The conductors and equipment for supplying electrical energy from the main or feeder, or from the transformer to the area that is served.

Serving, of Cable—A wrapping over the core of a cable before it is leaded, or over the lead if it is armored.

Shaded Pole—A single heavy conducting loop placed around one-half of a magnetic pole that develops an AC field, in order to induce an out-of-phase magnetic field.

Sheath, Cable — A protective covering (usually lead) applied to a cable.

Shell Core—A core for a transformer or reactor consisting of three legs, with the winding located on the center leg.

Short-Circuit — A fault path for current in a circuit that conducts excessive current; if the fault path has appreciable resistance, it is termed a leakage path.

Shunt — Denotes a parallel connection.

Sine Wave—Variation in accordance with simple-harmonic motion.

Sinusoidal — Having the form of a sine wave.

Sleeve, Splicing—Also termed "connector"; a metal sleeve (usually copper) which is slipped over and secured to the butted ends of conductors to make a

joint that provides good electrical connection.

Sleeve Wire — A circuit conductor connected to the sleeve of a plug or jack.

Sliding Contact — An adjustable contact arranged to slide mechanically over a resistive element, turns of a reactor, series of taps, or around the turns of a helix.

Snake—A steel wire or flat ribbon with a hook at one end, used to draw wires through conduit, etc.

Socket—A device for mechanical support of a device (such as a lamp) and for connection to the electrical supply.

Solenoid—A conducting helix with a comparatively small pitch; also applied to coaxial conducting helices.

Spark Coil — A l s o termed "ignition coil"; a step-up transformer designed to operate from a DC source via an interrupter that alternately makes and breaks the primary circuit.

Sparkover—A disruptive electrical discharge between the electrodes of a gap; generally used with reference to measurement of high voltage values with a gap having specified types and shape of electrodes.

Splice—Also termed "straight-through joint"; a series connection of a pair of conductors or cables.

Standard Cell—A highly precise source of DC voltage, also called a Weston cell; standard cells are used to check voltmeter calibration, and for highly precise measurement of DC voltage values.

Station, Automatic—A generating station or substation that is usually unattended and which performs its intended functions by an automatic sequence.

Surge — A transient variation in current and/or voltage at a given point in a circuit.

Switch—A device for making, breaking, or rearranging the connections of an electric circuit.

Symbol — A graphical representation of a circuit component; also, a letter or letters used to represent a component, electrical property, or circuit characteristic.

Tap—In a wiring installation, a T joint (Tee joint), Y joint, or multiple joint. Taps are made to resistors, inductors, transformers, etc.

Terminal — The terminating end(s) of an electrical device, source, or circuit, usually supplied with electrical connectors such as terminal screws, binding posts, tip jacks, snap connectors, soldering lugs, etc.

Three-Phase System—An AC system in which three sources energize three conductors, each of which provides a voltage that is 120° out of phase with the voltage in the adjacent conductor.

Tie Feeder—A feeder which is connected at both ends to sources of electrical energy. In an automatic station, a load may be connected between the two sources.

Time Delay—A specified period of time from the actuation of a control device to its operation of another device or circuit.

Tip, Plug—The contacting member at the end of a plug.

Torque —Mechanical twisting force.

Transfer Box — Also termed "pull box"; a box without a distribution panel containing branched or otherwise interconnected circuits.

Transformer — A device that operates by electromagnetic induction with a tapped winding, or two or more separate windings, usually on an iron core, for the purpose of stepping voltage or current up or down, for maximum power transfer, for isolation of the primary circuit from the secondary circuit, and in special designs for automatic regulation of voltage or current.

Transient—A nonrepetitive or arbitrarily timed electrical surge.

Transmission (AC) — Transfer of electrical energy from a source to a load, or to one or more stations for subsequent distribution.

Troughing—An open earthenware channel, wood, or plastic, in which cables are installed under a protective cover.

Tubing Electrical Metal(lic)— A thin-walled steel raceway of circular form with a corrosion-resistant coating for protection of wires or cables.

Tubing, Flexible—Also termed "loom"; a mechanical protection for electrical conductors; a flame-resistant and moisture-repellent circular tube of fibrous material.

Twin Cable—A cable consisting of two insulated and stranded conductors arranged in parallel runs and having a common insulating covering.

Underground Cable—A cable designed for installation below the surface of the ground, or for installation in an underground duct.

Ungrounded System — Also termed "insulated supply system"; an electrical system that "floats" above ground, or one that has only a very high-impedance conducting path to ground.

Unidirectional Current—A direct current or a pulsating direct current.

Units—Established values of physical properties used in measurement and calculation; for example, the volt unit, the ampere unit, the ampere-turn unit, the ohms unit, etc.

Value — The magnitude of a physical property expressed in terms of a reference unit, such as 117 volts, 60 Hz, 50 ohms, 3 henrys, etc.

VAR — Denotes volt-amperes reactive; the unit of imaginary power (reactive power).

Variable Component—A component that has a continuously controllable value, such as a rheostat, movable-core inductor, etc.

Vector — A graphical symbol for an alternating voltage or current, the length of which denotes the amplitude of the voltage or current, and the angle of which denotes the phase with respect to a reference phase.

Ventilated—A ventilated component is provided with means of air circulation for removal of heat, fumes, vapors, etc.

471

Vibrator—An electromechanical device that changes direct current into pulsating direct current (direct current with an AC component).

Volt—The unit of emf; one volt produces a current of one ampere in a resistance of one ohm.

Voltage — In a circuit, the greatest effective potential difference between a specified pair of circuit conductors.

Voltmeter—An instrument for measurement of voltage values.

Watt — The unit of electrical power, equal to the product of one volt and one ampere in DC values, or in rms AC values.

Watthour—A unit of electrical energy, equal to one watt operating for one hour.

Wattmeter—An instrument for measurement of electrical power.

Wave — An electrical undulation, basically of sinusoidal form.

Weatherproof — A conductor or device designed so that water, wind, or usual vapors will not impair its operation.

Wind Bracing—A system of bracing for securing the position of conductors or their supports to avoid the possibility of contact due to deflection by wind forces.

Wiper — An electrical contact arm.

Work—The product of force by the distance through which the force acts; work is numerically equal to energy.

Working Voltage—Also termed "closed-circuit voltage"; the terminal voltage of a source of electricity under a specified current demand; also, the rated voltage of an electrical component such as a capacitor.

XRay — An electromagnetic radiation with extremely short wavelength, capable of penetrating solid substances; used in industrial plants to check the perfection of device and component fabrication (detection of flaws). For other applications, refer to any standard electrical handbook.

Y Joint—A branch joint used to connect a conductor to a main conductor or cable for providing a branched current path.

Y Section — Also termed "T section"; an arrangement of three resistors, reactors, or impedances which are connected together at one end of each, with their other ends connected to individual circuits.

Zero-Adjuster — A machine screw provided under the window of a meter for bringing the pointer exactly to the zero mark on the scale.

Zero-Voltage Level — A horizontal line drawn through a waveform to indicate where the positive excursion falls to zero value, followed by the negative excursion. In a sine wave, the zero-voltage level is located half way between the positive peak and the negative peak.

Design Data for R.L.M.
Dome Reflectors (Direct-Lighting Luminaires)

The values in this table have been calculated using the following data:
Reflector efficiency (white bowl lamp) 65%
Allowance for depreciation 30 to 40%

These data apply to R.L.M. Dome reflectors equipped with white bowl Mazda lamps or slightly etched glass cover plate or glass ring when used with High Intensity Mercury Vapor Lamp. For clear or inside frosted Mazda lamps increase the average foot-candle values 10%.

Min. Mount. Ht. (Feet)	Approx. Luminaire Spacing (Feet)	Floor Area/Outlet (Sq.Ft.)	Room Factor	Average Foot-candles									
				Mazda Lamps								Mercury Lamps	
				100-W	150-W	200-W	300-W	500-W	750-W	1000-W	1500-W	250-W	400-W
8	8 x 8	60-70	A	8.5-12	15-19	21-27	35-44	56-75				42-56	
			B	6.5-8.5	11-15	16-21	26-35	43-56				32-42	
			C	5.0-6.5	8.5-11	12-16	19-26	33-43				25-32	
8½	9 x 9	70-90	A	7.5-10	12-17	17-23	27-38	47-65				35-48	
			B	5.5-7.5	8.5-12	13-17	20-27	35-47				26-35	
			C	4.0-5.5	6.5-8.5	9.5-13	15-20	26-35				19-26	
9½	10 x 10	90-110	A	6-8	10-13	13-18	22-30	35-50	52-73			26-37	57-80
			B	4.5-6.0	7.5-10	10-13	16-22	27-35	40-52			20-26	44-57
			C	3.5-4.5	5.5-7.5	7.5-10	12-16	21-27	31-40			16-20	34-44
10½	11 x 11	110-130	A	5.0-6.5	8-11	11-15	18-24	30-41	44-60			22-31	48-66
			B	4-5	6-8	8.5-11	13-18	23-30	34-44			17-22	37-48
			C	3-4	4.5-6.0	6.5-8.5	10-13	18-23	26-34			13-17	29-37
11	12 x 12	130-150	A	4.5-5.5	7-9	10-13	16-20	26-35	38-50	55-72		19-26	42-55
			B	3.5-4.5	5.5-7.0	7.5-10	12-16	20-26	29-38	42-55		15-19	32-42
			C	2.5-3.5	4.0-5.5	5.5-7.5	9-12	15-20	22-29	32-42		11-15	24-32
11½	13 x 13	150-180	A		6-8	8-11	13-18	22-30	33-44	44-62		17-23	36-48
			B		4.5-6.0	6-8	10-13	17-22	25-33	34-44		13-17	27-36
			C		3.5-4.5	4.5-6.0	7.5-10	13-17	19-25	26-34		9.5-13	21-27
12	14 x 14	180-210	A		5.5-6.5	7.5-9.0	11-15	18-25	27-36	39-52		14-19	30-40
			B		4.0-5.5	5.5-7.5	8.5-11	14-18	21-27	30-39		11-14	23-30
			C		3-4	4.0-5.5	6.5-8.5	11-14	16-21	23-30		8-11	20-23
12½	15 x 15	210-240	A		4.5-5.5	6-8	9-13	17-22	23-31	34-44	53-70	13-16	25-34
			B		3.5-4.5	4.5-6.0	7-9	13-17	18-23	26-34	40-53	10-13	20-25
			C		2.5-3.5	3.5-4.5	5.5-7.0	9.5-13	14-18	20-26	31-40	7-10	16-20
13½	16 x 16	240-270	A		3.5-5.0	5-7	8.5-11	14-19	21-27	30-39	47-61	11-14	23-30
			B		2.5-3.5	4-5	6.5-8.5	11-14	16-21	23-30	36-47	8-11	18-23
			C		2.0-2.5	3-4	5.0-6.5	8.5-11	12-16	18-23	28-36	6.5-8.0	13-18
14	17 x 17	270-300	A			5-6	8-10	13-17	18-24	27-35	42-54	10-13	20-26
			B			4-5	6-8	9.5-13	14-18	21-27	32-42	7-10	16-20
			C			3-4	4.5-6.0	7.5-9.5	11-14	16-21	25-32	5.5-7.0	12-16
15	18 x 18	300-340	A			4.5-5.5	6.5-9.0	11-15	17-22	23-31	37-49	8-11	19-24
			B			3.5-4.5	5.0-6.5	9-11	13-17	18-23	28-37	7-8	15-19
			C			2.5-3.5	4-5	7-9	10-13	14-18	22-28	5-7	11-15
15½	19 x 19	340-380	A			3.5-5.0	6-8	10-13	15-19	22-27	33-43	7.5-10	17-21
			B			2.5-3.5	4.5-6.0	8-10	12-15	17-22	26-33	6.0-7.5	13-17
			C			2.0-2.5	3.5-4.5	6-8	9-12	13-17	20-26	4.5-6.0	10-13
16	20 x 20	380-420	A				5.5-7.0	9-12	13-17	18-25	30-39	7-9	15-19
			B				4.0-5.5	7-9	10-13	14-18	23-30	5-7	11-15
			C				3-4	5.5-7.0	8-10	11-14	18-23	4-5	9-11
17	21 x 21	420-460	A				5.0-6.5	8.5-11	13-16	17-22	26-35	6.5-8.0	15-18
			B				4-5	6.5-8.5	10-13	13-17	20-26	5.0-6.5	11-15
			C				3-4	5.0-6.5	7.5-10	10-13	16-20	4-5	8-11
18	22 x 22	460-500	A				4.5-6.0	8-10	11-14	16-20	24-32	6.0-7.5	12-16
			B				3.5-4.5	6-8	8.5-11	12-16	19-24	4.5-6.0	9.5-12
			C				2.5-3.5	4.5-6.0	6.5-8.5	9.5-12	15-19	3.5-4.5	7.0-9.5
19	23 x 23	500-550	A				4.0-5.5	6.5-9.0	10-13	14-19	23-29	5-7	11-15
			B				3-4	5.0-6.5	8-10	11-14	18-23	4-5	9-11
			C				2.5-3.0	4.5	6-8	8.5-11	14-18	3-4	6.5-9.0

Design Data for Semidirect Lighting Luminaires

The values in this table have been calculated using the following data: Luminaire efficiency 75 to 90% Allowance for depreciation 30%	Prismatic Glass		Opal Glass					

Min. Mount. Ht. (Feet)	Approx. Luminaire Spacing (Feet)	Floor Area/Outlet Sq.Ft.	Room Factor	Average Foot-candles							
				100-W	150-W	200-W	300-W	500-W	750-W	1000-W	1500-W
8	8 x 8	60-70	C	7.5-11	12-18	19-26	29-42	50-72			
			D	5.0-7.5	9-12	13-19	20-29	34-50			
			E	3.5-5.0	6-9	8.5-13	14-20	23-34			
8½	9 x 9	70-90	C	6.5-9.5	11-16	15-22	24-36	40-62	56-90		
			D	4.5-6.5	7-11	9.5-15	16-24	27-40	38-56		
			E	3.0-4.5	4.5-7.0	6.5-9.5	11-16	18-27	26-38		
9½	10 x 10	90-110	C	5.0-7.5	7.5-12	12-17	19-28	32-48	46-70	65-100	
			D	3.5-5.0	5.0-7.5	8-12	13-19	22-32	31-46	44-65	
			E	2.5-3.5	3.5-5.0	5.5-8.0	8.5-13	15-22	21-31	30-44	
10½	11 x 11	110-130	C	4.5-6.0	6.5-10	9.5-14	16-23	26-39	38-57	54-81	
			D	3.0-4.5	4.5-6.5	6.5-9.5	11-16	18-26	26-38	37-54	
			E	2-3	3.0-4.5	4.5-6.5	7.5-11	12-18	18-26	25-37	
11	12 x 12	130-150	C		6.0-8.5	9-12	14-20	23-33	35-48	47-68	75-108
			D		4-6	6-9	9.5-14	16-23	24-35	32-47	51-75
			E		3-4	4-6	6.5-9.5	11-16	16-24	22-32	35-51
11½	13 x 13	150-180	C		5.0-7.5	6.5-10	12-17	19-29	28-42	38-59	63-93
			D		3.5-5.0	4.5-6.5	8-12	13-19	19-28	26-38	43-63
			E		2.5-3.5	3.0-4.5	5.5-8.0	9-13	13-19	18-26	29-43
12	14 x 14	180-210	C			6.0-8.5*	9.5-14	16-24	23-35	35-49	54-78
			D			4.5-6.0	6.5-9.5	11-16	16-23	24-35	37-54
			E			3.0-4.5	4.5-6.5	7.5-11	11-16	16-24	25-37
12½	15 x 15	210-240	C			5.0-7.5	9-12	14-20	19-30	29-42	47-67
			D			3.5-5.0	6-9	10-14		20-29	32-47
			E			2.5-3.5	4-6	7-10	9.5-13	14-20	22-32
13½	16 x 16	240-270	C				7.5-11	12-18	18-26	26-37	41-58
			D				5.0-7.5	8.5-12	12-18	18-26	28-41
			E				3.5-5.0	8.0-8.5	8.5-12	12-18	19-28
14	17 x 17	270-300	C				6.5-9.5	11-16	16-23	23-33	37-52
			D				4.5-6.5	7.5-11	11-16	16-23	25-37
			E				3.0-4.5	5.5-7.5	8-11	11-16	17-25
15	18 x 18	300-340	C				6.0-8.5	10-14	15-21	21-30	32-47
			D				4.5-6.0	7-10	10-15	15-21	22-32
			E				3.0-4.5	5-7	7-10	10-15	15-28
15½	19 x 19	340-380	C				5.0-7.5	9-13	12-18	19-26	29-41
			D				3.5-5.0	6.5-9.0	9-12	13-19	20-29
			E				2.5-3.5	4.5-6.5	6-9	8.5-13	14-20
16	20 x 20	380-420	C					7.5-11	11-16	16-23	26-37
			D					5.5-7.5	8-11	11-16	18-26
			E					4.0-5.5	5.5-8.0	8-11	12-18
17	21 x 21	420-460	C					7-10	11-15	15-21	23-33
			D					5-7	7.5-11	10-15	16-23
			E					3.5-5.0	5.0-7.5	7-10	11-16
18	22 x 22	460-500	C					6.5-9.5		13-19	21-30
			D					4.5-6.5	6.5-9.5	9.5-13	15-21
			E					3.0-4.5	4.5-6.5	6.5-9.5	10-15
19	23 x 23	500-550	C					6.0-8.5	9-12	12-18	19-28
			D					4.5-6.0	6-9	9-12	13-19
			E					3.0-4.5	4-8	6-9	6.5-13

Design Data for Semi-Indirect Luminaires

The values in this table have been calculated using the following data: Luminaire efficiency 65 to 85% Allowance for depreciation 30%					Opal Glass	Prismatic Glass	Opal Bowl				

Min. Ceil. Ht. (Feet)	Min. Hngr. Lgt. (Feet)	Approx. Luminaire Spacing (Feet)	Floor Area/Outlet (Sq.Ft.)	Room Factor	Average Foot-candles							
					100-W	150-W	200-W	300-W	500-W	750-W	1000-W	1500-W
8	½	8 x 8	60-70	C	6.0-9.5	9.5-16	14-22	23-36	39-60			
				D	4-6	6.0-9.5	8.5-14	15-23	25-39			
				E	2.5-4.0	4-6	5.5-8.5	9-15	16-25			
8½	1	9 x 9	70-90	C	5-8	8-13	12-19	18-30	31-52	44-75		
				D	3-5	4.5-8.0	7-12	11-18	19-31	28-44		
				E	2-3	3.0-4.5	4.5-7.0	7-11	12-19	18-28		
9½	1½	10 x 10	90-110	C	4-6	6.5-10	9-15	15-24	24-36	36-58		
				D	2.5-4.0	4.0-6.5	5.5-9.0	9.5-15	16-24	23-36		
				E	1.5-2.5	2.5-4.0	3.5-5.5	6.0-9.5	10-16	15-23		
10½	2	11 x 11	110-130	C		5.0-8.5	8-12	12-19	21-33	30-48	44-68	
				D		3-5	4.5-8.0	8-12	13-21	19-30	28-44	
				E		2-3	3.0-4.5	5-8	8.5-13	12-19	18-28	
11	2	12 x 12	130-150	C			6-10	10-16	17-28	26-40	36-57	
				D			4-6	7-10	11-17	17-26	23-38	
				E			2.5-4.0	4.5-7.0	7.5-11	11-17	15-23	
11½	2	13 x 13	150-180	C			5.0-8.5	8.5-14	15-24	22-35	31-50	48-78
				D			3-5	5.5-8.5	9.5-15	14-22	20-31	31-48
				E			2-3	3.5-5.5	6.0-9.5	9-14	13-20	20-31
12	2½	14 x 14	180-210	C			5.0-7.5	8-12	13-20	19-29	27-41	41-65
				D			3-5	4.5-8.0	8-13	12-19	17-27	26-41
				E			2-3	3.0-4.5	5-8	7.5-12	11-17	17-26
12½	2½	15 x 15	210-240	C				6-10	11-17	16-25	23-35	36-56
				D				4-6	7-11	10-16	15-23	25-36
				E				2.5-4.0	4.5-7.0	6.5-10	9.5-15	15-23
13½	3	16 x 16	240-270	C				6-9	9.5-15	14-22	20-31	31-49
				D				4-6	6.0-9.5	9.5-14	13-20	20-31
				E				2.5-4.0	4-6	6.0-9.5	8.5-13	13-20
14	3	17 x 17	270-300	C				4.5-8.0	8.5-13	12-19	17-28	29-43
				D				3.0-4.5	5.5-8.5	8.5-12	11-17	19-29
				E				2-3	3.5-5.5	5.5-8.5	7.5-11	12-19
15	3	18 x 18	300-340	C					8.5-12	11-17	16-25	26-39
				D					5.5-8.5	7-11	10-16	17-26
				E					3.5-5.5	4.5-7.0	6.5-10	11-17
15½	3	19 x 19	340-380	C					7-11	10-15	14-22	23-35
				D					4.5-7.0	6.5-10	9.5-14	15-23
				E					3.0-4.5	4.0-6.5	6.0-9.5	9.5-15
16	4	20 x 20	380-420	C					6.0-9.5	9.5-14	13-20	20-31
				D					4-6	6.0-9.5	8.5-13	13-20
				E					2.5-4.0	4-6	5.5-8.5	8.5-13
17	4	21 x 21	420-460	C					5.5-8.5	8.5-13	12-18	19-28
				D					4.0-5.5	5.5-8.5	8-12	12-19
				E					2.5-4.0	3.5-5.5	5-8	8-12
18	4	22 x 22	460-500	C					5-8	7-11	11-16	16-25
				D					3-5	4.5-7.0	7-11	11-16
				E					2-3	3.0-4.5	4.5-7.0	7-11
19	4	23 x 23	500-550	C						7-10	9.5-15	15-23
				D						4.5-7.0	6.0-9.5	10-15
				E						3.0-4.5	4-6	6.5-10

Design Data for Indirect Lighting Luminaires

The values in this table have been calculated using the following data:
Luminaire efficiency 65 to 85%
Allowance for depreciation 30%

Opaque Reflector   Luminous Bowl

Min. Ceil. Ht. (Feet)	Min. Hngr. Lgt. (Feet)	Approx. Luminaire Spacing (Feet)	Floor Area/ Outlet (Sq.Ft.)	Room Factor	100-W	150-W	200-W	300-W	500-W	750-W	1000-W	1500-W
8	½	8 x 8	60-70	C	5.0-8.5	8-14	11-19	20-32	33-54	49-78		
				D	3-5	5-8	7.5-11	12-20	20-33	30-49		
				E	2-3	3-5	4.5-7.5	7-12	12-20	18-30		
8½	1	9 x 9	70-90	C	4-7	7-12	10-17	16-27	28-46	41-67		
				D	2.5-4.0	4-7	5.5-10	9-16	17-28	23-41		
				E	1.5-2.5	2.5-4.5	3.5-5.5	5.5-9.0	9.5-17	14-23		
9½	1½	10 x 10	90-110	C			8-13	13-21	22-36	32-52	43-74	
				D			5-8	7.5-13	13-22	19-32	26-43	
				E			3-5	4.5-7.5	8-13	11-19	16-26	
10½	2	11 x 11	110-130	C			6.5-11	11-17	18-29	26-42	38-60	67-95
				D			4.0-6.5	6.5-11	11-18	16-26	23-38	41-67
				E			2.5-4.0	4.0-6.5	6.5-11	9.5-16	14-23	25-41
11	2	12 x 12	130-150	C			6-9	10-15	15-25	22-36	33-51	50-80
				D			3.5-6.0	6-10	9-15	13-22	20-33	34-50
				E			2.0-3.5	3.5-6.0	5.5-9.0	8-13	12-20	21-34
11½	2	13 x 13	150-180	C			4.5-8.0	8-13	13-21	18-31	26-44	43-69
				D			2.5-4.5	5-8	7.5-13	11-18	16-26	28-43
				E			1.5-2.5	3-5	4.5-7.5	7-11	10-16	18-28
12	2½	14 x 14	180-210	C				6.5-11	11-18	16-26	23-37	39-58
				D				4.0-6.5	6.5-11	10-16	14-23	24-39
				E				2.5-4.0	4.0-6.5	6-10	8.5-14	15-24
12½	2½	15 x 15	210-240	C				5.5-9.0	9-15	13-22	21-32	31-50
				D				3.5-5.5	5.5-9.0	8-13	13-21	19-31
				E				2.0-3.5	3.5-5.5	5-8	7.5-13	12-19
13½	3	16 x 16	240-270	C				5.5-8.0	8-13	12-19	18-28	26-43
				D				3.5-5.5	5-8	7.5-12	11-18	16-26
				E				2.0-3.5	3-5	4.5-7.5	8.5-11	10-16
14	3	17 x 17	270-300	C					8-12	11-17	16-25	24-39
				D					5-8	6.5-11	10-16	15-24
				E					3-5	4.0-6.5	6-10	9-15
15	3	18 x 18	300-340	C					6.5-11	9-16	13-22	21-35
				D					4.0-6.5	5.5-9.0	8-13	13-21
				E					2.3-4.0	3.5-5.5	5-8	8-13
15½	3	19 x 19	340-380	C					6.0-9.5	8.5-14	12-20	20-31
				D					4-6	5.0-8.5	7.5-12	12-20
				E					2.5-4.0	3-5	4.5-7.5	7.5-12
16	4	20 x 20	380-420	C					5.0-8.5	8-12	11-18	18-28
				D					3.5-5.5	5-8	6.5-11	11-18
				E					2.0-3.5	3-5	4.0-6.5	6.5-11
17	4	21 x 21	420-460	C						6.5-11	10-16	16-25
				D						4.0-6.5	6.5-10	10-16
				E						2.5-4.0	4.0-6.5	6-10
18	4	22 x 22	460-500	C						6.5-10	9.0-14	14-23
				D						4.0-6.5	5.5-9.0	9-14
				E						2.5-4.0	3.5-5.5	5.5-9.0
19	4	23 x 23	500-550	C						6.0-9.5	8-13	13-21
				D						3.5-6.0	5-8	8-13
				E						2.0-3.5	3-5	5-8

Index